INNOVATIVE FINANCE APPROACHES FOR ADDRESSING RIVER BASIN POLLUTION

COMBATING AQUATIC BIODIVERSITY LOSS IN SOUTHEAST ASIA

JUNE 2024

ACGF
ASEAN CATALYTIC GREEN FINANCE FACILITY

ADB

© 2024 Asian Development Bank
6 ADB Avenue, Mandaluyong City, 1550 Metro Manila, Philippines
Tel +63 2 8632 4444; Fax +63 2 8636 2444
www.adb.org

Some rights reserved. Published in 2024.

ISBN 978-92-9270-755-2 (print); 978-92-9270-756-9 (PDF); 978-92-9270-757-6 (e-book)
Publication Stock No. SPR240254-2
DOI: http://dx.doi.org/10.22617/SPR240254-2

The views expressed in this publication are those of the authors and do not necessarily reflect the views and policies of the Asian Development Bank (ADB) or its Board of Governors or the governments they represent.

ADB does not guarantee the accuracy of the data included in this publication and accepts no responsibility for any consequence of their use. The mention of specific companies or products of manufacturers does not imply that they are endorsed or recommended by ADB in preference to others of a similar nature that are not mentioned.

By making any designation of or reference to a particular territory or geographic area in this document, ADB does not intend to make any judgments as to the legal or other status of any territory or area.

Corrigenda to ADB publications may be found at http://www.adb.org/publications/corrigenda.

Notes:
In this publication, "$" refers to United States dollars, "€" refers to euro, "¥" refers to yen, "B" refers to baht, "CNY" refers to yuan, and "D" refers to dong.

ADB recognizes "China" as the People's Republic of China; "Hong Kong" as Hong Kong, China; "Korea" as the Republic of Korea; "Laos" as the Lao People's Democratic Republic; and "Vietnam" as Viet Nam.

On the cover: Blue boats transporting tourists going to Phong Nha cave. The pier was funded by ADB's Greater Mekong Subregion Sustainable Tourism Development Project. Training programs also gave the local people opportunities to work in the tourism sector in and around the National Park (photo by Ariel Javellana).

Contents

Tables, Figures, and Boxes

Boxes

Author Biographies

Anouj Mehta, country director of the Thailand Resident Mission (TRM), Asian Development Bank (ADB)

Anouj leads the full range of work in the planning, implementation, and supervision of the TRM vision, goals, strategies, and work plan in support of overall ADB goals in the context of its engagement with an upper-middle income country. He has been a core member of the ADB team developing the Nature Solutions Finance Hub, which was launched at the United Nations Climate Change Conference in December 2023. He is the former head of the ASEAN Catalytic Green Finance facility, a pioneering regional facility under the ASEAN Infrastructure Fund, aimed at green recovery in the region with 13 partners and $1.9 billion in financing support. Before his current role, he established and managed the Green Finance Hub (formerly the Innovation Hub) for the Southeast Asia Regional Department of ADB, focusing on green and innovative finance. He has also previously led one of ADB's pioneering public–private partnerships initiatives in India. Before joining ADB, he was an investment banker and chartered accountant at JP Morgan Chase and PwC in London.

Manuela Cavaccini Cataldo, Green Finance Initiatives coordinator, TRM, ADB

Manuela supports the TRM country director in coordinating and monitoring the roll-out of four planned and/or underway initiatives (GSS+ Green Bonds, Nature Solutions Finance Hub, Zero Source Pollution, and OneADB Green Blue initiative) for Thailand and the region. In addition, she works closely with the Southeast Asia Department Green Finance Hub and other ADB departments, including the Climate Change and Sustainable Department, the ASEAN Catalytic Green Finance Facility, and the Sectors Group. In 2022, as the knowledge management specialist, Manuela developed a filter mechanism to leverage regional technical assistance better and assess their additional value. Before joining ADB, she acquired over 10 years of international experience in business development, project, and partner management in both the private and the public sectors.

Bingxun Seng, partner, EY

Bingxun is a partner at EY and leads the Economics Advisory practice in Singapore. He is an experienced economist with 15 years of experience in providing strategic policy advice for governments, private entities, multilateral development entities, and nonprofits across Southeast Asia. He has an extensive portfolio of work across the international development space, particularly across topics such as trade integration, industry development, infrastructure development, circular economy and the green transition, and public finance management. For example, he has collaborated with ADB in proposing reform recommendations to support the recovery of key sectors in Southeast Asia after the coronavirus disease (COVID-19), advised on the economic and financial feasibility of major infrastructure projects in the region, and has also advised Cambodia, the Lao People's Democratic Republic, Myanmar, and Viet Nam on strategies to narrow the development gaps with their ASEAN counterparts.

Oliver Redrup, associate partner, EY

Oliver is an associate partner based in the Singapore Infrastructure Advisory team, with over 17 years of experience throughout ASEAN, the United Kingdom (UK), the Middle East, and the United States advising government and private sector clients on the development, financing, procurement, and delivery of complex, large-scale infrastructure projects. Oliver has worked extensively with multilateral development banks and on donor-funded infrastructure programs to improve the delivery of infrastructure projects and to apply innovative financing structures such as blended financing. He holds an MBA from Manchester Business School and a BSc in Economics from the University of Warwick.

Yvonne Xie, associate director, EY

Yvonne is an associate director at EY with over a decade of experience in economics consulting across urban development and infrastructure sectors. She has led multiple engagements for government, multilateral organizations, and private sector clients in Asia, and her expertise is in conducting industry, market feasibility, and economic planning studies. Her project portfolio includes transport and utilities, commercial mixed-use developments, integrated tourism developments, industrial parks, and new city developments. She holds an MBA from the Hong Kong University of Science and Technology and a bachelor's degree in environmental engineering from the National University of Singapore.

Terence Lee, senior associate, EY

Terence is a part of EY Singapore's Economic Advisory team. He has nearly 4 years of experience working on various economic modeling and research projects. He has supported clients in quantifying the economic impacts of their operations and has recently developed demand projections for a proposed regional port. He holds a master's degree in management (University of Cambridge, UK) and a bachelor's degree in economics (the London School of Economics, UK).

Acknowledgments

Innovative Finance Approaches for Addressing River Basin Pollution: Combating Aquatic Biodiversity Loss in Southeast Asia was developed by the Asian Development Bank (ADB) with support from the technical assistance Policy Advice for COVID-19 Economic Recovery in Southeast Asia (Phase 2) and the ASEAN Catalytic Green Finance Facility.

The report was led under the supervision of Anouj Mehta, country director of the ADB Thailand Resident Mission supported by Manuela Cavaccini Cataldo, Green Finance Initiatives coordinator. The study is a collaboration between ADB and EY Corporate Advisors Pte. Ltd., led by Bingxun Seng, with support from Oliver Redrup, Yvonne Xie, and Terence Lee. Overall production coordination of the report was managed by Marina Lopez Andrich and Roykaew Nitithanprapas. Editing was done by Melanie Kelleher; design and layout by Rodel Valenzuela; proofreading by Mrinmoy Gogoi; and page proofchecking by Jess Alfonso Macaset.

Methodologically, this report's insights were developed based on an extensive review of professional and academic literature and supplemented with expert interviews.

The authors wish to acknowledge the invaluable advice granted by Mova Al'Afghani, Silvia Cardascia, Keoduangchai Keokhamphui, Thammarat Koottatep, Pinida Leelapanang, Ly Van Loi, Kongmeng Ly, Paul Pavelic, Reni Suwarso, Farhan Dzakwan Taufik, Au Shion Yee, and Ming Li Yong in developing this report. The authors are also grateful for the valuable inputs and discussions with ADB colleagues Jiangfeng Zhang, Ancha Srinivasan, Michael Rattinger, Isao Endo and Geoffrey Wilson.

Abbreviations

ADB	Asian Development Bank
AIT	Asian Institute of Technology
ASEAN	Association of Southeast Asian Nations
ATS	algal turf scrubbers
DFAT	Department of Foreign Affairs (Australia)
EPR	extended producer responsibility
EPT	Environmental Protection Tax (People's Republic of China)
EU	European Union
GDP	gross domestic product
GEF	Global Environment Facility
GEMStat	global freshwater quality database
GMS	Greater Mekong Subregion
IBRD	International Bank for Reconstruction and Development
ICMA	International Capital Market Association
ICPDR	International Commission for the Protection of the Danube River
ICPR	International Commission for the Protection of the Rhine
IFC	International Finance Corporation
IRF	International River Foundation
JBIC	Japan Bank for International Cooperation
Lao PDR	Lao People's Democratic Republic
LMB	Lower Mekong Basin
MCC	Millennium Challenge Corporation
MRC	Mekong River Commission
MSW	municipal solid waste
NBS	nature-based solutions
NGO	nongovernment organization
NIB	Nordic Investment Bank
NSF Hub	Nature Solutions Finance Hub
OCBC	Oversea-Chinese Banking Corporation Limited
ODA	official development assistance
OECD	Organisation for Economic Co-operation and Development
PES	payment for ecosystem services
PET	polyethylene terephthalate
PPP	public–private partnership
PRC	People's Republic of China

PWRF	Philippine Water Revolving Fund
SDG	Sustainable Development Goal
SeyCCAT	Seychelles' Conservation and Climate Adaptation Trust
TMDL	total maximum daily load
TTB	TMBThanachart Bank
UK	United Kingdom
UNEP	United Nations Environment Programme
US	United States
USAID	United States Agency for International Development
WWTP	wastewater treatment plant

Executive Summary

Southeast Asia is a globally recognized hotspot of biological diversity and endemism, but it is also among the most biotically threatened regions on the planet. Habitats in the region are endangered by a range of causes, each of which raises the likelihood of species extinction in a variety of habitats. Rivers are the lifeblood of the land, people, and economies they support, and these ecosystems are under increasing pressure from climate change and pollution, putting at risk the water sources that support billions of people.

The Problem

The problem of water crisis majorly driven by river pollution is particularly severe in the rapidly growing Southeast Asian region, resulting in deteriorating water quality; hampering aquatic and human life; and negatively impacting economic, social, and environmental conditions in the region. As river pollution levels increase, the delicate balance of ecosystems is disrupted, leading to the decline and loss of numerous aquatic species. This loss of biodiversity has cascading effects on the health of the planet, impacting water quality, food security, and human well-being.

Southeast Asia is home to numerous rivers that traverse the region, nourishing its landscapes and playing a vital role in the livelihoods of millions of people. The Mekong, Hong (Red), Chao Phraya, and Citarum rivers are often cited as the most important rivers of the region because of their significant geological features, biodiversity, and economic benefits. All of these rivers are facing a serious threat related to water pollution.

At the heart of this region lies the Mekong River. Agricultural activity and urban settlements across the value chain are major sources of the delta's polluted waters. The water quality of the Hong River—also known as the Red River—is degrading because of densely populated areas, rapid economic growth, and large industrial zones surrounding the delta. The Chao Phraya River, which flows through the heart of Thailand—including Bangkok—plays a critical role in irrigating rice paddies and supporting a variety of aquatic life. The concentration of agricultural activity and population pressures have turned Bangkok into a pollution hotspot along the Chao Phraya. The Citarum River in Indonesia—often referred to as one of the most polluted rivers in the world—is also affected significantly by human activity and industrial development, including textile factories.

After introducing the subject (Chapter 1), this report begins by framing the problem of river basin pollution (Chapter 2). It starts by examining the extent of the river pollution in the region, followed by an assessment of the extent of water pollution in the four major rivers in Southeast Asia and Europe's major rivers, analyzing water quality based on three core indicators: dissolved oxygen, total phosphorous, and total nitrogen (section 2.2).

The report highlights key drivers and anthropogenic activities responsible for river pollution in Southeast Asia such as population growth, rapid urbanization, industrialization, agricultural practices, municipal and plastic waste, the increasing use of plastics, and inadequate wastewater treatment infrastructure, among which plastic pollution poses a severe threat to aquatic biodiversity (section 2.3). Within Asia, Southeast Asia has emerged as a hotspot for plastic pollution, which comprises more than half of the top 10 countries in the world contributing to plastic leakage that eventually reaches the oceans (section 2.4).

The report briefly explains how waste in water leads to a detrimental impact on biodiversity, climate change, and the population's livelihood and health. Waste in water (i) imperils the rich aquatic biodiversity by disrupting ecosystems and endangering various species; (ii) exacerbates climate change by releasing greenhouse gases and undermining the region's carbon sequestration capacity; and (iii) jeopardizes the livelihoods and health of local communities, as contaminated water sources contribute to waterborne illnesses, diminish agricultural productivity, and threaten the social and economic well-being of those who rely on these vital river systems (Chapter 3).

The Solution

The relevance of resilient river basins in meeting the Sustainable Development Goals—particularly poverty alleviation, water security, ecological protection, and climate change adaptation and mitigation—is becoming more widely recognized. Combating river water pollution, achieving water security, and equitable basin-wide use in Southeast Asia can be done by applying integrated approaches and managing river basins as unified ecological units. It also necessitates the development of regulatory frameworks and policy incentives to strengthen basin-wide planning and decision-making processes, as well as utilizing innovative financing mechanisms to leverage public sector finance, encourage private sector engagement, and advance sustainable development.

The report demonstrates how river pollution problem can be addressed by adopting nature-based solutions (NBS), policy measures, institutional arrangements, and financing mechanisms (Chapter 4):

(i) **NBS** plays a significant role in fighting water-related hazards and their negative consequences, especially in the face of climate change that will intensify existing constraints (section 4.1). The report explores how NBS can be harnessed in three ways: (a) using and protecting natural ecosystems, (b) restoring ecosystems, and (c) creating new ecosystems.

(ii) Two types of **policy measures**—command and control policies and economic incentives or market-based approaches—can encourage stakeholders to reduce pollution, adopt more sustainable practices, and protect river ecosystems while also benefiting from cost savings, financial rewards, and market advantages for environmentally responsible actions (section 4.2).

(iii) **Institutional arrangements** promote the practical implementation of these policies by coordinating government agencies, developing technical and administrative ability, engaging communities, and raising public awareness (section 4.3). Policy measures and institutional arrangements work in tandem to produce a symbiotic interaction for putting other approaches presented in this report into action. For example, having adequate institutional arrangements ensures the successful implementation of NBS, the proper execution of various legislative measures, and the effectiveness of employing innovative financing tools that facilitate the implementation of these solutions.

(iv) The study highlights that Southeast Asia faces a burgeoning infrastructure financing gap for ensuring sustainable economic expansion. It outlines significant obstacles—such as an overall lack of funding, inadequate policy development, disorganized institutional frameworks, a lack of innovative financing products, poorly designed projects, and insufficient experience of the private sector—as the key challenges faced in Southeast Asia in obtaining sufficient funding to meet the Sustainable Development Goals. Various **innovative financing mechanisms** and instruments are introduced for consideration in supporting pollution control and improvement in water quality (section 4.4). Examples of these innovative financing include blue bonds, public–private partnerships, and blended finance, which could involve a mix of public and private funding sources.

The report concludes by providing a holistic review of the range of approaches applicable to combating river pollution, leading to recommendations for the next steps to be taken by stakeholders in the Greater Mekong Subregion to catalyze the implementation of projects that address river pollution challenges, along with discussion about pilot projects that can ultimately lead to the development of robust and scalable solutions (Chapter 5).

1 INTRODUCTION

Greater Mekong Subregion Route 3—Thailand.
The last section of Route 3 will be built across the
Mekong River, between Houey Xai in the Lao People's
Democratic Republic and Chiang Khong in Thailand
within the next few years (photo by Ariel Javellana/ADB).

Southeast Asia, which covers only 4% of the earth's land area, is a region rich in endemic biodiversity. At least 6 of the world's 25 biodiversity hotspots—the areas of the world that contain an exceptional concentration of species and are exceptionally endangered—are found in Southeast Asia. In addition to this biodiversity, the region boasts an extraordinary rate of species discovery, with more than 2,216 new species discovered during 1997–2014.[1] Nature's economic contribution to the Association of Southeast Asian Nations (ASEAN) member states is estimated to be about $2.2 trillion, with the figure expected to rise if nations further act on conservation.[2] However, the region's biodiversity is under serious threat, with estimates revealing that 13%–42% of species will be lost in the region by 2100, half of which will be global extinctions.[3]

Southeast Asia is particularly vulnerable to aquatic biodiversity loss with many of its 665 million population living on long coastlines. Rivers are the lifeblood of the land, people, and economies they support, and poor water quality and increasing abstraction of water are putting pressure on the environment. Water quality in Asia has deteriorated significantly, with pollution increasing in 50% of major rivers during 1990–2010, salinity increasing by more than one-third, and 80% of wastewater being discharged into waterways without adequate treatment.[4] Nine countries in Southeast Asia—Cambodia, Indonesia, the Lao People's Democratic Republic (Lao PDR), Malaysia, Myanmar, Philippines, Thailand, Timor-Leste, and Viet Nam—remained in the lower two "insecure" categories of the five band national water security index published by the Asian Development Bank (ADB) in 2020.[5]

River basin pollution in Southeast Asia is a particular challenge with the quality of water dramatically impacted—whether in cities or rural areas—by four main pollutants: (i) municipal waste (including from informal settlements), (ii) toxic industrial effluents, (iii) agricultural run-off such as chemicals and fertilizers, and (iv) plastics. This "waste in water" is causing the following negative triple impacts:

(i)   degradation of the critical water biosphere, including freshwater and the oceans;

(ii) an increase in the emission of greenhouse gases from microbial activity that converts pollutants into these gases; and

(iii) a reduction in the availability of clean water for lives and livelihoods, leading to inequity in economic growth.

The vital role that rivers in the Greater Mekong Subregion (GMS) play in supporting its large population and important ecosystem is increasingly being impaired. The GMS is a natural economic area in Asia bound by the Mekong River, covering 2.6 million square kilometers (km^2) and a combined population of about 326 million.[6] More than 70 million people depend partly or entirely on the Mekong River as a source of income and life, but this very important ecosystem is under tremendous stress from climate change, hydropower development, pollutant runoff from farms and municipalities, and growing use of watercourses as a means of rubbish disposal.

---

[1]   A. Hughes. 2017. *Even as more new species are found, Southeast Asia is in the grip of a biodiversity crisis.* The Conversation. 5 January.
[2]   Academy of Sciences Malaysia. 2022. *The Nexus of Biodiversity Conservation and Sustainable Socioeconomic Development in Southeast Asia.*
[3]   Food and Agriculture Organization of the United Nations (FAO). n.d. *Asia-Pacific Forests and Forestry to 2020.* Forest Policy Brief 01.
[4]   ADB. 2020. *Asian Water Development Outlook 2020.*
[5]   Effective 1 February 2021, ADB placed a temporary hold on sovereign project disbursements and new contracts in Myanmar.
[6]   ADB. *Greater Mekong Subregion (GMS).*

This report seeks to raise awareness of the importance of protecting the rivers of the region, and the basins from which their water is drawn, given the triple impacts. It also highlights key challenges and solutions to catalyze projects for river pollution reduction and control.

It begins by presenting an overview of the major rivers of Southeast Asia, key sources of pollution, and the pollution impact. It then reviews a range of interventions both within and beyond the region covering nature-based solutions, policy measures, institutional arrangements, and financing mechanisms, with deeper coverage of innovative financing options. Finally, this report offers suggestions on the next steps to be taken to implement innovative yet practical projects to address the river pollution problem within the GMS area.

2 AN OVERVIEW OF THE POLLUTION STATUS OF MAJOR RIVERS IN SOUTHEAST ASIA

Central Mekong Delta Region Connectivity Project. A corner of Can Tho city is taken from above (photo by Viet Tuan/ADB).

The water crisis in Southeast Asia has been a pressing concern because of a combination of factors such as population growth, rapid urbanization, industrialization, and climate change. The resulting water-related issues such as deteriorating water quality and frequent water shortages are intensifying the water crisis, hampering aquatic and human life, and negatively impacting economic, social, and environmental conditions in the region. The following points underscore water pollution concerns in the region:

(i) About 110 million people in Southeast Asia live without access to safe water.[7]

(ii) Only about one-third of all the wastewater in Asia is treated, with the lowest treatment rates in South Asia (7%) and Southeast Asia (14%).[8] Southeast Asia faces many challenges related to wastewater treatment, including inadequate infrastructure, insufficient funding, and regulatory enforcement issues.

(iii) Water pollution has significantly affected both the quality and the availability of fish. About 37% of aquatic species are affected by pollution in Southeast Asia.[9]

(iv) Southeast Asia accounts for 80% of global aquaculture and 60% of the world's capture fisheries.[10] Water pollution has a detrimental impact on both fish production and consumption, creating a situation of excess demand. For instance, the region is expected to witness a 36% increase in fish consumption by 2030.[11]

(v) As of 2020, about 356 million people in Southeast Asia were living in urban areas. The urban population is expected to increase by 70 million in 2050, further intensifying the stress on water resources in the region.[12]

Controlling water pollution in Southeast Asia—with a focus on river pollution—is of paramount importance to protect human health, aquatic ecosystems, and the environment. Controlling and addressing river pollution in the region is a crucial step requiring comprehensive and coordinated efforts by multiple stakeholders—including governments, local communities, industries, and international partners—to improve the water quality and environmental health of rivers in the region.

This chapter introduces several major rivers in Southeast Asia that stand out for their geological, ecological, and economic importance. It then provides insight into their health by mapping their river basin pollution status against standards designed to protect freshwater ecosystems as these ecosystems provide a vital supply of water and food, harbor exceptional biodiversity (including more fish species than have been discovered in the world's oceans),[13] and are linked to several Sustainable Development Goals (SDGs).[14]

To place these results in a global context, their pollution status is compared to that of several major European rivers. These European rivers once faced severe pollution challenges but have since benefited from pollution control measures, providing good case studies for river cleanups in Southeast Asia. With the region's rivers functioning more like drains and sewers, this chapter provides a review of the major sources of pollution afflicting the major rivers of Southeast Asia and their toxicity and concludes with a special note on plastic pollution within the region.

[7] ChinaDialogue. 2020. *The Right to Safe Water in Southeast Asia.*
[8] Development Asia. 2021. *Economic Tools to Help Manage Asia's High Demand for Water Resources.*
[9] ADB. 2023. *Zero Source Pollution Initiative (ZSPI) Fighting Biodiversity Loss: Tackling Waste in Water.*
[10] ADB. 2022. *Breaking the Waves: Kickstarting the Global Sustainable Blue Economy in Southeast Asia.*
[11] World Bank. 2013. *FISH TO 2030 Prospects for Fisheries and Aquaculture.*
[12] ADB. 2022. *Technical Assistance for Smart and Livable Cities in Southeast Asia.* Manila.
[13] United Nations Enviroment Progamme (UNEP). 2020. *Freshwater Ecosystems Tool Enables SDG Reporting to Continue Despite COVID-19.*
[14] ADB. 2020. *The Health of Asia's Rivers Lies Both in the Cities and on the Farms.*

2.1 Major Rivers in Southeast Asia

Southeast Asia hosts dozens of rivers that meander both within and across country borders. The Mekong, Hong (Red), Chao Phraya, and Citarum rivers are often cited as the region's major rivers because of their outstanding geological features, biodiversity, and economic contributions. Key highlights of these major rivers are available in Table 1.

Table 1: Key Features of Major Rivers of Southeast Asia

Mekong	
Length	4,800 km
Basin area	810,000 km²
Countries traversed	Cambodia, Lao People's Democratic Republic, Myanmar, Thailand, Viet Nam
National capitals served	Phnom Penh, Vientiane
Geology	Longest river in Southeast Asia, flowing from the Himalayas and through five Southeast Asian nations
Biodiversity	Rich in biodiversity (e.g., 1,200 fish species) and home to ecological zones of national, regional, and global significance (e.g., wetlands with Ramsar site status)
Economic features	The Lower Mekong Basin (LMB)—which rests within mainland Southeast Asia—supports the lives and livelihoods of 52 million people. Key sectors supported include agriculture and the world's largest inland fishery with an annual turnover of $1.4–$3.9 billion.
Hong (Red)	
Length	1,200 km
Basin area	156,500 km²
Countries traversed	Viet Nam
National capitals served	Ha Noi
Geology	Largest river in Northern Viet Nam, flowing through Ha Noi and 25 provinces
Biodiversity	The Red River Delta's coastal areas are home to a complex system of various vegetation types. In particular, the mangrove and intertidal habitats form wetlands of high biodiversity that also serve as globally important sites for migratory birds.
Economic features	Supports a population of 26 million people, with a total agricultural area of nearly 1.09 million hectares
Chao Phraya	
Length	365 km
Basin area	159,000 km²
Countries traversed	Thailand
National capitals served	Bangkok
Geology	Flows south through Thailand's fertile central plain, situated between Ayutthaya and the Gulf of Thailand
Biodiversity	Hosts about 280 fish species, including three of the largest freshwater fish in the world: the critically endangered giant barb, giant pangasius, and giant freshwater stingray
Economic features	Home to 57% of Thailand's population, including the capital Bangkok. The Basin area generates 66% of the country's GDP
Citarum	
Length	297 km
Basin area	13,000 km²
Countries traversed	Indonesia
National capitals served	Jakarta

continued on next page

Table 1 *continued*

Citarum	
Geology	Longest river in Indonesia's West Java province. Source of water for the country's largest reservoir, the Jatiluhur
Biodiversity	Contains 160 plant species and is also home to a sizeable number of mammals, birds, reptiles, and fish species that have been classified as threatened
Economic features	Home to 5 million people in the river basin

GDP = gross domestic product, km = kilometer, km^2 = square kilometer.
Notes:
[1] More details on these rivers' features are available in Appendix 1. Countries traversed include those within Southeast Asia only.
[2] Effective 1 February 2021, Asian Development Bank placed a temporary hold on sovereign project disbursements and new contracts in Myanmar.
Source: Asian Development Bank.

2.2 Pollution Status of Major Rivers of Southeast Asia against Global Comparators

2.2.1 Global Comparators: Major Rivers of Europe

Key rivers traversing continental Europe were identified as global comparators given that river pollution management measures from these countries could constitute global best practices. These comprise the Danube, Elbe, Rhine, and Seine that were once severely polluted but have been moving toward recovery. Table 2 provides some additional information about each river turnaround story. More details on these rivers' pollution recovery highlights are available in Appendix 2.

Table 2: General Background of Global Comparators and Their Pollution Recovery Highlights

Danube	
Length	2,900 km
Basin area	801,500 km^2
Countries traversed	Austria, Bulgaria, Croatia, Germany, Hungary, Moldova, Romania, Serbia, Slovakia, Ukraine
History of river pollution and recovery highlights	Severely polluted by the mid-1980s because of population growth and industrialization, approximately 80% of the Danube's wetlands and floodplains disappeared. Is exhibiting clear signs of recovery, for example, with nitrogen emissions having decreased by 20% during 2000—2015.
Elbe	
Length	1,100 km
Basin area	148,000 km^2
Countries traversed	Austria, Czechia, Germany, Poland
History of river pollution and recovery highlights	By the end of the 1980s, the Elbe was considered one of Europe's most polluted rivers, the waters carrying a toxic mix of pollutants including pentachlorophenol (a highly toxic chemical compound). However, water quality has substantially improved, as reflected by the considerable increase in the river's number of fish species.
Rhine	
Length	1,320 km
Basin area	225,000 km^2
Countries traversed	Austria, France, Germany, the Netherlands, Switzerland, Liechtenstein,
History of river pollution and recovery highlights	On the brink of death in the 1960s, industrial pollution caused the extinction of the Rhine salmon and the deaths of even the hardiest of fish and eel species. Measures to improve water quality have led to a marked recovery of the Rhine's biodiversity. Almost all the Rhine's fish species have returned.

continued on next page

Table 2 *continued*

Seine	
Length	750 km
Basin area	76,000 km²
Countries traversed	France
History of river pollution and recovery highlights	Pollution from industrial and municipal waste has historically plagued the Seine. By the 1960s, only three species of fish were recorded in Paris. Considerable improvements in water quality have led to the reintroduction of swimming after a century, as well as an increase in the number and size of fish.

km = kilometer, km² = square kilometer.

Source: Asian Development Bank.

2.2.2 Pollution Status Mapping against Global Comparators

At its core, this mapping exercise utilizes the *Framework for Freshwater Ecosystem Management* developed by the United Nations Environment Programme (UNEP) to assess the pollution status of these Southeast Asian river basins alongside their global comparators. This framework was chosen as it has been applied to map the pollution status of freshwater bodies globally via the UNEP global freshwater quality database (GEMStat) program.[15]

Under UNEP GEMStat global mapping, water quality is assessed against standards established to protect freshwater ecosystems using four globally relevant core parameters: (i) dissolved oxygen concentration, (ii) total phosphorus, (iii) total nitrogen, and (iv) pH (measure of acidity and alkalinity). Each parameter has three water quality classification categories, with the categorization approach being based on a survey of internationally and nationally established standards (Table 3).[16]

Table 3: United Nations Environment Programme GEMStat Water Quality Classification System

Indicators	Class A—Natural	Class B to D—Non-Natural Disturbed	Class E—Seriously Disturbed
Dissolved Oxygen (DO) Concentration (mg/L)	7.3–10.9	3–7.3 or 10.9–13.6	<3 or >13.6
Total Phosphorus (TP) (ug/L)	<20	20–190	>190
Total Nitrogen (TN) (ug/L)	<700	700–2,500	>2,500
pH	6.5–9.0	5–6.5	<5

ug/L= micrograms per liter, mg/L = milligrams per liter, pH = measure of acidity and alkalinity.

Note: The classification system provides total phosphorus and total nitrogen standards for "rivers and streams" and "lakes and reservoirs" separately. This study considers only the former given its greater relevance in the context of this analysis.

Source: United Nations Enviroment Progamme GEMStat Program. (n.d.) Water Quality Indicators.

---

[15] UNEP. 2023. *GEMStat Water Quality Indicators*.
[16] UNEP. 2017. *A Framework for Freshwater Ecosystem Management. Volume 2: Technical Guide for Classification and Target-Setting*.

Table 4 presents the results of this study's pollution status mapping.[17] As Table 4 reveals, the European comparators' water quality tends to be better. However, rivers across Europe were once heavily polluted through the 20th century because of industrial, agricultural, and urban expansion. European governments subsequently took action to address the matter from the 1970s onward, beginning with the issuance of the first water directives. The European Union (EU) has since worked to create an effective and coherent water policy for the region. The introduction of the EU Urban Wastewater Treatment Directive (1991), the Water Framework Directive (2000), and River Basin Management Plans (2010, 2016) further aligned efforts to improve water quality and aquatic ecosystem health.[18] These efforts by the EU—together with national legislation—have resulted in remarkable improvements.

Table 4: Pollution Status of the River Basins of Southeast Asia against European Comparators

Indicators	Southeast Asia				European Comparators			
	Mekong	Hong (Red)	Chao Phraya	Citarum	Danube	Rhine	Elbe	Seine
Dissolved Oxygen (DO) Concentration (mg/L)	5.1	2.6	2.0	0.8	8.3	10.4	8.8	10.3
Total Phosphorus (TP) (ug/L)	170.0	320.0	1,100.0	350.0	70.0	77.0	120.0	90.0
Total Nitrogen (TN) (ug/L)	3,040.0	4,100.0	21,800.0	26,000.0	1,300.0	2,600.0	3,190.0	3,350.0
pH	7.5	6.9	7.2	7.8	8.0	8.1	7.7	7.9
Data Years	2019	2019	2017–2019	2015, 2017, 2019, 2020, 2023	2019	2019	2019	2015

ug/L = micrograms per liter, mg/L = milligrams per liter, pH = measure of acidity and alkalinity.

Legend, Water Quality Classification System: Blue (Natural), Orange (Non-Natural Disturbed), Red (Seriously Disturbed)

Source: Literature review, EY analysis.

In contrast, the water quality crisis in Southeast Asia has intensified over the past few decades, with Table 4 showing that the key Southeast Asian rivers assessed are mostly seriously disturbed across the three core parameters (dissolved oxygen, total phosphorus, and total nitrogen). Box 1 presents the example of the most polluted river in the world and the actions of the government to rehabilitate it.

---

[17] Pollution data was gathered from a review of publicly available sources. While effort has been made to incorporate water quality data across these rivers' basins, data limitations restrict the survey to key sub-areas in some cases (e.g., Hong [Red] River Delta). Caution should thus be exercised when interpreting these mapping results. They should not be interpreted as a comparison of pollution status across the entirety of each river's basin (detailed notes on areas surveyed by river basin are available in Appendix 3). In addition, a conservative approach has been adopted to assign pollution status. Where multiple mean readings across a river basin are available (e.g., a mean across a mainstream, and another mean across a tributary), pollution status has been assigned based on the poorest mean reading available.

[18] European Environment Agency (EEA). 2020. *Freshwater quality*.

Box 1: The Citarum—The World's Most Polluted River

The Citarum River is the longest river in West Java, Indonesia, stretching 297 kilometers before reaching the Java Sea. As of 2023, more than 5 million people live in the river basin, and the river provides water, electricity, and irrigation to more than 25 million people.

The Citarum River in West Java, Indonesia

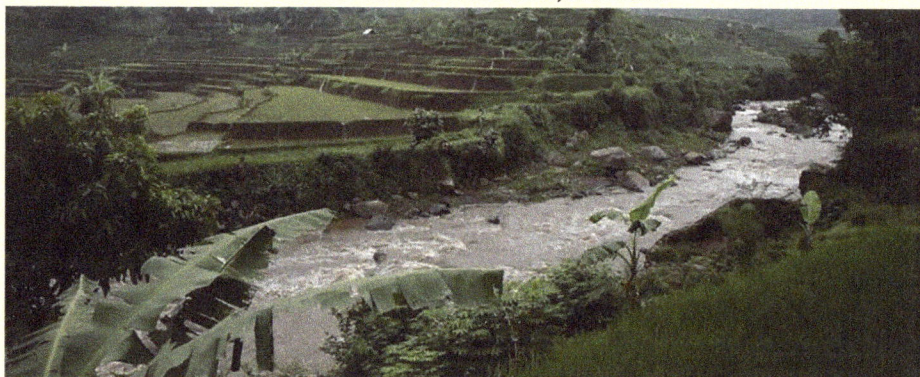

Source: ADB. 2014. *Tackling Pollution in Indonesia's Citarum River Basin.*

Billed as the world's most polluted river, the Citarum contains a mix of rubbish, chemical waste, and dead fish that come together to produce a dense, repelling odor. Pollution of the river's water is severe, with studies documenting levels of toxic lead 1,000 times worse than US drinking water standards. As a result of being laden with pollutants, people suffer from various skin diseases such as scabies and infections, and aquatic life has starkly diminished, with 60% of Citarum fish species disappearing since 2008.

Recognizing that heavy pollution of the Citarum requires addressing, the Government of Indonesia has embarked on a program to make the river's water drinkable by 2025.[a, b] The government estimated that the *Citarum Harum* (Fragrant Citarum) program would require up to Rp35 trillion ($2.34 billion) to be spent on numerous infrastructure projects and the day-to-day dredging, cleaning, and patrolling.[c] In 2008, ADB committed to providing Indonesia with a $500 million loan to finance a wide-ranging cleanup and rehabilitation plan for the Citarum River basin.[d] It is estimated that a clean Citarum River will bring more than $280 million per year in benefits to Indonesia through resource recovery, reduced costs of drinking water production, increased yields from fish farming, enhanced real estate, and associated opportunities for tourism and biodiversity.[e]

The situation has started to improve gradually since the Indonesian government began to clamp down on factories that violated environmental laws and relocate illegal settlements that disposed of solid waste, detergents, and feces into the river and its tributaries. As of 2023, there are spots in the river where the water is safe enough to swim. Scavengers who once hunted for plastic cups and bottles became fishers instead. There remains much work to be done as activists and experts have seen various administrations trying to clean the river only for their programs to lose progress or struggle with funding. It has also been observed that much of Fragrant Citarum's success rests on efforts to continuously monitor and search for evidence on environmental law violations as well as the constant work to pick up rubbish polluting the river. There are calls for more pollution prevention strategies and less reliance on monitoring (footnote c).

ADB = Asian Development Bank, Rp = Indonesian rupiah, US = United States.
Sources:
[a] *The Guardian.* 2020. Rotten River: Life on One of the World's Most Polluted Waterways. Photo Essay. 2 November.
[b] *The Diplomat.* 2018. Indonesia's Citarum: The World's Most Polluted River. April.
[c] ChannelNewsAsia. 2021. *Signs of life after Indonesia's polluted Citarum river gets a clean up—but for how long?*
[d] ADB. 2014. *Cleaning up Indonesia's Citarum Basin.*
[e] ADB. 2013. *Downstream Impacts of Water Pollution in the Upper Citarum River, West Java, Indonesia.*

2.3 Major Sources of Pollution and Their Toxicity

There are two primary sources of water pollution—point and nonpoint—from which pollutants enter water bodies. These two differ in their characteristics and methods of pollution discharge. Point sources are well-defined and single identifiable locations such as oil wells, wastewater treatment plants, power plants, and underground coal mines. Nonpoint sources are unidentifiable sources of pollution that are not associated with specific discharge points. Some examples of non point sources are irrigation, agriculture sediment run-off, and agrochemical discharge.

The quality of water in Southeast Asian rivers is significantly affected by four main pollutants:

(i) **Municipal waste**—This includes waste discarded by households, businesses, and institutions within urban areas, as well as from rural areas and informal settlements. In Southeast Asia, the urban population is projected to grow by 8.5% from 2015 to 2030 while the economy is projected to grow at a compound annual growth rate of 4.9% from 2022 to 2030.[19] Rapid urbanization, population growth, and economic growth will continue to place significant pressure on waste management infrastructure.

(ii) **Industrial waste**—Industries discharge toxic industrial effluents comprising chemicals, heavy metals, and radioactive materials into water bodies, affecting aquatic life, human health, and the environment.

(iii) **Agricultural waste**—Increasing agriculture, livestock, and aquaculture activities are responsible for deteriorating water quality. Agriculture farms often discharge large quantities of agrochemicals, organic matter, drug residues, sediments, and saline drainage into water bodies.

(iv) **Plastic waste**—Microplastics, discarded plastic items, and single-use plastics are posing an increasing threat to marine life. More than 80% of plastic that enters the oceans comes from rivers in Asia.[20]

River pollution in Southeast Asia is a complex issue with multiple contributing factors in different countries. This section seeks to highlight pollution hotspots within the Mekong, Hong (Red), Chao Phraya, and Citarum River basins and to identify key sources of pollution within them. Tackling both point and nonpoint source pollution will remain key to restoring the health of these rivers.

2.3.1 Mekong River

Pollution hotspots have developed in the Lower Mekong Basin, including the Mekong Delta

The Lower Mekong Basin (LMB) has experienced noticeable water pollution, especially near the Lao People's Democratic Republic's capital Vientiane; the Sesan, Srepok, and Sekong rivers (the 3S river system); the Tonle Sap Lake system, and the Mekong Delta in southern Viet Nam.[21]

Agricultural activity across the value chain is a major source of the delta's polluted waters

Agricultural activity is presently the Mekong Delta's largest source of pollution. Aquaculture is a key contributor as 456 million cubic meters (m³) of sediment and waste are discharged annually from aquaculture farms.[22]

---

19   UNhabitat.org. 2022. *ASEAN Sustainable Urbanisation Report.*
20   OurWorldinData. 2022. *Ocean plastics: How much do rich countries contribute by shipping their waste overseas?*
21   R. Sor et al. 2021. Water Quality Degradation in the Lower Mekong Basin. *Water.* 13(11). p. 1555; R. Chea et al. 2016. Evidence of Water Quality Degradation in Lower Mekong Basin Revealed by Self-Organizing Map. *PLoS One.* 11(1). e0145527.
22   Institute for Environmental Science and Development (VESDEC). 2017. *Water Pollution in the Mekong Delta: Sources, Present, Future, Ecological Impacts and Mitigation.*

These include Pangasius fish farms that have experienced rapid growth.[23] Apart from aquaculture farms, seafood processing plants higher up the value chain are another source of pollution. For instance, shrimp processing factories in the Hòa Trung Industrial Zone have been suspected of discharging wastewater into the environment, polluting water sources.[24]

Animal farming discharges 14.6 million m³ of wastewater and 8.2 million tons of solid waste annually (footnote 22). These include a range of harmful compounds, including animal waste, antibiotics, and hormones.[25]

Crop agriculture pollutes the delta's waters as well. The discharge of excavated soil sediments increases the need for these particles to be removed before use. In addition, the sediments have the potential to destroy habitats of the smallest organisms, causing declines in fish populations higher up the food chain.[26] Other pollutants include pesticides, the use of which has dramatically increased in Viet Nam since its transformation in the 1990s.[27] Crop processing plants are another source of river pollution. For instance, a local rice processing firm was suspected of releasing pollutive wastewater that lowered dissolved oxygen levels and increased nitrate, and phosphate levels, causing mass fish death.[28]

Rural and urban settlements are key contributors to the delta's water quality woes

Another major source of pollution is the development of settlements and the attendant discharge of municipal waste. As of 2023, there are more than 17.5 million inhabitants in the Mekong Delta, of which more than 70% live in rural areas. This leads to the intensive use of the delta's rivers and canals for daily living, as residents prefer to settle along the canal banks, discharging wastewater and solid waste directly into canals. As for urban areas, water quality is susceptible to urbanization, which has increased five fold in the LMB over the last 2 decades, primarily in the delta (as well as the Mekong's 3S river system, and around Tonle Sap Lake). Urbanization leads to residential development from which household waste is discharged to rivers.[29] Annually—and within the delta—102 million m³ of wastewater and 606,000 tons of solid waste (including 4,000 tons of medical waste) are generated. Compounding the matter is the fact that treatment rates vary: more than 85% of solid wastes and 100% of medical wastes are collected and treated but only 30% of effluents are treated (footnote 22). In summary, the delta's rural and urban settlements are a significant source of water quality deterioration (footnote 27).

2.3.2   Hong (Red) River

Rapid economic growth in the Red River Delta has turned the area into a pollution hotspot

The Red River—the largest river in northern Viet Nam—serves as the main water source for production and human activities in the Red River Delta region. Cities and provinces in the Red River Delta—for example, Ha Noi, Nam Dinh, and Ha Nam—have experienced rapid economic growth with the development of large urban, industrial, and agricultural zones.[30] This has taken a toll on the health of the delta's waters.

---

23   P. Anh et al. 2010. Water Pollution by Pangasius Production in the Mekong Delta, Vietnam: Causes and Options for Control. *Aquaculture Research*. 42(1). pp. 108–128.

24   *Viet Nam News*. 2017. Pollution Causes Mass Fish Deaths in Cà Mau.

25   F. Renaud and C. Kuenzer. 2012. *The Mekong Delta System: Interdisciplinary Analyses of a River Delta*. Springer Science and Business Media.

26   R. Chea et al. 2016. Evidence of Water Quality Degradation in Lower Mekong Basin Revealed by Self-Organizing Map. *PLoS One*. 11(1). e0145527; Mid-America Regional Council. What Is Sediment Pollution?

27   R. Chea et al. 2016. Evidence of Water Quality Degradation in Lower Mekong Basin Revealed by Self-Organizing Map. *PLoS One*. 11(1). e0145527.

28   R. Sor et al. 2021. Water Quality Degradation in the Lower Mekong Basin. *Water*. 13(11), 1555.

29   VESDEC. 2017. *Water Pollution in the Mekong Delta: Sources, Present, Future, Ecological Impacts and Mitigation*; P. Anh et al. 2010. Water Pollution by Pangasius Production in the Mekong Delta, Vietnam: Causes and Options for Control. *Aquaculture Research*. 42(1). pp. 108–128.

30   T. Tham et al. 2022. Assessment of Some Water Quality Parameters in the Red River Downstream, Vietnam by Combining Field Monitoring and Remote Sensing Method. *Environmental Science and Pollution Research*. 1(13).

Densely populated areas are degrading the quality of the delta's waters

Densely populated areas contribute to the delta's tremendous domestic wastewater problem. Domestic wastewater in the delta accounts for 23% of the nation's total because of the area's high population density.[31] In this regard, the population density influence on water quality is reflected in the higher partial pressure of carbon dioxide in waters near densely populated (and rapidly growing) Ha Noi compared to other parts of the delta as higher volumes of sewage discharged trigger the production of carbon dioxide when organic matter contained in sewage decomposes.[32]

Agricultural activity—a key sector—is another key source of pollution

The delta's position as the second most important rice-producing area in Viet Nam (accounting for 20% of national production) places significant stress on the quality of the delta's waters.[33] The excessive use by farmers of chemical fertilizers pollutes waters with nitrogen, threatening the environment and human health. Most rice farmers apply chemical fertilizers above the recommended rates with the hope of increased crop yield. However, fertilizer use efficiency is low at high levels of application resulting in excess fertilizer entering the soil and water. As such, more than half of the nitrogen load in paddy fields stems from chemical fertilizers.[34]

Large industrial zones and smaller craft villages both contribute to the pollution challenge

The Red River Delta is characterized by several large industrial zones and many craft villages that contaminate the delta's waters with industrial effluent and wastewater.[35] Craft villages (household enterprises and small and medium-sized enterprises)—which have experienced accelerating growth—are a prominent feature of the delta. Often located near the delta's waters, they are characterized using old, inefficient technologies, and have limited capacity to undertake environmental protection measures such as the installation of wastewater treatment systems. For instance, Ha Noi's 1,350 craft villages discharge up to 57 million m^3 of untreated wastewater annually. Although not as significant a source of pollution compared to large-scale industrial zones, they remain a concern as high concentrations of craft villages can cause accumulative, localized, and downstream pollution.[36]

2.3.3   Chao Phraya River

Concentration of industrial activity and population pressures have turned Bangkok into a pollution hotspot along the Chao Phraya

The concentration of Thailand's industrial activity around Bangkok and the industry discharge of industrial effluents is a key contributor to the city's declining water quality.[37] Almost one-third of the country's output from industrial parks is produced in Bangkok, and the capital is home to many small-scale factories, with larger plants located near the port. These industries are mainly involved in the textiles sector, food processing, electronic equipment assembly, and production of building materials.

31   Clean Currents Coalition. *Citarum River, Indonesia.*
32   S. McGowan et al. 2023. *Water Quality in Rivers of the Red River Delta.* University of Nottingham Asia Research Institute.
33   H. Nguyen et al. 2016. Seasonal Variability of Faecal Indicator Bacteria Numbers and Die-off Rates in the Red River Basin, North Viet Nam. *Scientific Reports.* 6(1). p. 21644.
34   *Viet Nam News.* 2017. Pollution Causes Mass Fish Deaths in Cà Mau.
35   F. Renaud and C. Kuenzer. 2012. *The Mekong Delta System: Interdisciplinary Analyses of a River Delta.* Springer Science and Business Media; H. Long and M. Yabe. 2011. The Impact of Environmental Factors on the Productivity and Efficiency of Rice Production—A Study in Vietnam's Red River Delta. *European Journal of Social Sciences.* 26. pp. 218–230.
36   T. Dang and T. Tran. 2020. Rural Industrialization and Environmental Governance Challenges in the Red River Delta, Vietnam. *The Journal of Environment and Development.* 29(4). pp. 420–448.
37   N. Singkran et al. 2020. BOD Load Analysis and Management Improvement for the Chao Phraya River Basin, Thailand. *Environmental Monitoring and Assessment.* 192. pp. 1–14.

Wastewater from communities is another source of degraded water quality around Bangkok.[38] A cause of this has been limited affordable housing in urban areas. This has encouraged the growth of densely populated informal settlements along the capital's canals, with these communities utilizing the waters they live by as sewers (footnote 33).

Agricultural activity along the Chao Phraya is also a cause for concern

Agriculture—crop farming—is another cause for concern. Crop agriculture releases agrochemical runoffs with intensive use (and subsequent release) of chemical fertilizers being a reason for polluted waters. By way of background, the use of chemical fertilizers in Thailand increased more than 100 times from the 1960s to about 2004. Despite the massive increase in chemical fertilizer use, yields of rice and maize only doubled once in 45 years. This indicates a tremendous loss of fertilizers in the environment because of poor management.[39] Given this backdrop, crop farming along the Chao Phraya remains a pollution concern.

2.3.4 Citarum River

Industrial activity—by the textile sector in particular—significantly harms the Citarum's waters

Industrial activity contributes significantly to the Citarum River's degraded water quality. The region is dominated by the textile industry that has not taken sufficient measures to treat industrial wastewater containing toxic chemicals and heavy metals. In 2017, only 10% of the 3,200 textile factories had wastewater treatment facilities.[40] While enforcement action has been taken against errant textile factories, factories continue to release untreated wastewater by evading authorities on patrol.[41] In addition, large factories related to other sectors such as food and shoes are a source of pollution, alongside micro, small, and medium-sized enterprises that line the river (footnote 40).

Direct discharge of wastewater by settlements is another major source of pollution

Domestic wastewater from settlements such as households, stalls, and restaurants is directly disposed of into the river body, releasing pollutive organic material.[42] In addition, detergent from washing has been identified as a contributor to the river's elevated phosphate levels (footnote 40). Unfortunately, municipal wastewater's contribution to the Citarum's woes is set to grow as the amount of waste produced by those living along the river is not slowing down.[43]

2.4 Plastic Pollution

Plastic pollution is a significant environmental concern globally. Four hundred sixty million tons of plastic is produced worldwide every year, a figure expected to triple by 2060.[44] Most of the plastic waste is not treated, resulting in almost 150 million tons of plastic waste accumulating in the world's rivers and oceans.[45] Annually, up to 23 million tons of plastic waste leak into aquatic ecosystems, polluting lakes, rivers, and seas. This amount is expected to triple

---

[38] S. McGowan et al. 2023. *Water Quality in Rivers of the Red River Delta.* University of Nottingham Asia Research Institute.

[39] R. Tirado et al. 2008. *Use of Agrochemicals in Thailand and Its Consequences for the Environment.* Devon: Greenpeace Research Laboratories.

[40] A. Utami et al. 2020. The Pollutant Load in Downstream Segment of Citarum River, Indonesia. *International Journal of Scientific and Technological Research* 9(01). p. 3506.

[41] *The Jakarta Post.* 2020. West Java Court Declares Textile Company Guilty of Polluting Citarum River; *Channel News Asia.* 2021. Signs of Life after Indonesia's Polluted Citarum River gets a Clean Up—But for how Long?

[42] A. Utami et al. 2020. The Pollutant Load in Downstream Segment of Citarum River, Indonesia. *International Journal of Scientific and Technological Research* 9(01). p. 3506; R. Yokosawa and T. Mizunoya. 2022. Improving Water Quality in the Citarum River through Economic Policy Approaches. *Sustainability.* 14(9). p. 5038.

[43] Channel News Asia. 2021. Signs of Life after Indonesia's Polluted Citarum River Gets a Clean Up—But for How Long?

[44] Organisation for Economic Co-operation and Development (OECD). 2022. *Plastic Pollution is Growing Relentlessly as Waste Management and Recycling Fall Short.*

[45] ADB. 2022. *Stopping Plastic Waste Leakage into Oceans.*

by 2040.[46] Despite government and industry commitment to address the plastic challenge, annual plastic flow into the oceans is only expected to decline annually by 7% in 2040.[47]

Southeast Asia is a significant contributor to the world's plastic pollution crisis

Asia is a significant contributor to plastic pollution in water bodies. Fifteen of the world's top 20 polluting rivers flow through Asia.[48] Ninety-five percent of the global load of plastics is transported into the sea through 10 rivers, 8 of which are in Asia: the Amur, Ganges, Hai, Indus, Mekong, Pearl, Yangtze, and Yellow rivers (footnote 45). Southeast Asia has emerged as a hotspot for plastic pollution because of rapid urbanization, a rising middle class, and inadequate infrastructure for waste management.[49] Countries in the region now comprise more than half of the top 10 countries contributing to plastic leakage that eventually reaches the oceans (Figure 1).[50]

Microplastics are of particular concern because of their harmful effects and difficulties in removing them

Plastic pollution comprises plastic debris as well as microplastics (less than 5 millimeters).[51] This special note focuses on the sources and impacts of microplastics considering their harmful effects, as well as difficulties with removing them from the environment.[52]

2.4.1　Sources of Microplastic Pollution

Human activities in urban areas and settlements

At a broad level, plastic pollution arises from human activities in urban areas and settlements. For example, a source of plastic pollution is the Mekong River in Phnom Penh—where only an estimated 17% of the city's plastic waste is collected—with plastic from the city subsequently reaching the delta in Viet Nam.[53] The Chao Phraya River also carries plastic pollution from the millions who reside along the river—including the capital Bangkok—and the upper provinces.[54] In the Citarum River, an abundance of microplastics has been associated with the area's population density (18.6 million people live along the stream), and industrial activity, with about 3,000 industries discharging plastic-laden wastewater into streams.[55]

Domestic, industrial, and agricultural activity

Unmanaged and managed waste disposal from domestic sources both contribute to microplastics in rivers, highlighting difficulties with their flow. Inefficiencies in solid waste collection result in the direct disposal of plastic into canals and rivers,[56] where microplastics are mainly generated from photodegradation and water movement.[57]

---

[46]　UN. 2023. *Solutions to Plastic Pollution.*

[47]　Pewtrust. 2020. *Breaking the Plastic Wave: Top Findings for Preventing Plastic Pollution.*

[48]　The Circulate Initiative. 2022. 2021 Annual Report.

[49]　World Bank. 2022. *Turning the Tide on Plastic Pollution through Regional Collaboration in Southeast Asia.*

[50]　PlasticBank. 2023. *Which Country Is the Most Accountable for Ocean Plastic?*

[51]　K. Afkarina et al. 2020. Distribution and Environmental Risk of Microplastics Pollution in Freshwater of Citarum Watershed. *E3S Web of Conferences.* 211. p. 03012. EDP Sciences.

[52]　N. Le et al. 2023. Microplastics in the Surface Sediment of the Main Red River Estuary. *Vietnam Journal of Earth Sciences.* 45(1). pp. 19–32.

[53]　C. Haberstroh et al. 2021. Plastic Transport in a Complex Confluence of the Mekong River in Cambodia. *Environmental Research Letters.* 16(9). p. 095009.

[54]　T. Anh et al. 2020. Microplastics Contamination in a High Population Density Area of the Chao Phraya River, Bangkok. *Journal of Engineering and Technological Sciences.* 52(4); P. Jendanklang et al. 2023. Distribution and Flux Assessment of Microplastic Debris in the Middle and Lower Chao. *Journal of Water and Health.*

[55]　E. Sembiring et al. 2020. The Presence of Microplastics in Water, Sediment, and Milkfish (Chanos chanos) at the Downstream Area of Citarum River, Indonesia. *Water, Air, and Soil Pollution.* 231. pp. 1–14.

[56]　K. Ounjai et al. 2022. Assessment of Microplastic Contamination in the Urban Lower Chao Phraya River. *Journal of Water and Health.* 20(8). pp. 1243–1254.

[57]　P. Chanpiwat et al. 2021. Abundance and Characteristics of Microplastics in Freshwater and Treated Tap Water in Bangkok, Thailand. *Environmental Monitoring and Assessment.* 193. pp. 1–15.

Figure 1: World's Top 10 Ocean Plastic Contributors in 2021

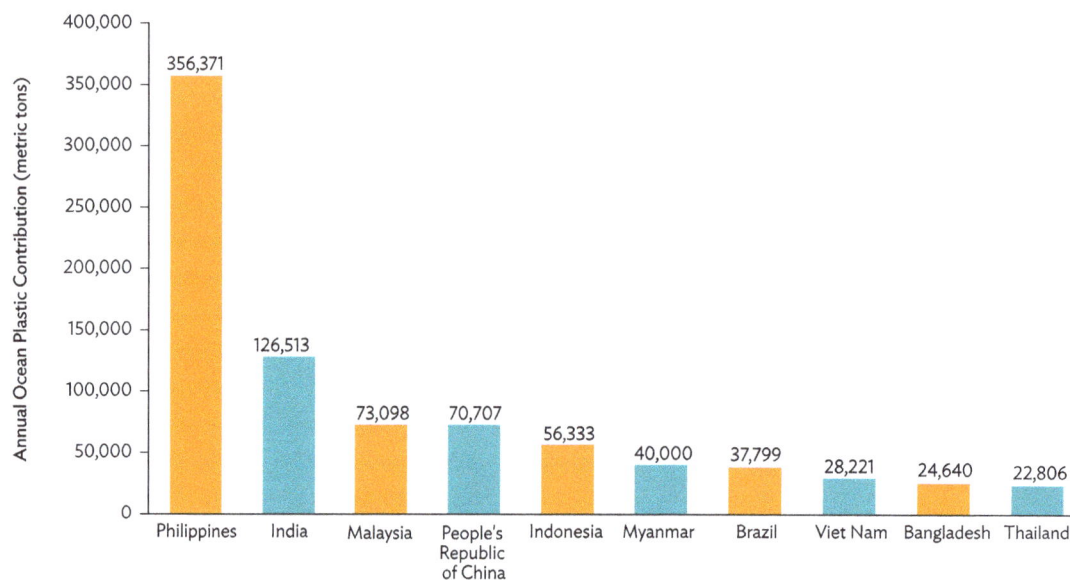

Source: PlasticBank. 2023. *Which Country Is the Most Accountable for Ocean Plastic?*

In cases where solid waste is transferred to landfills, plastic waste may likewise degrade and flow into rivers through runoff.[58] In addition, domestic wastewater contains microplastics—typically from the washing and laundering of synthetic clothes[59]—and cosmetic and beauty products that contain microbeads.[60] Moreover, given their minute size, sewage treatment plants may not be able to completely filter microplastics (footnote 50).

Industrial effluent is usually untreated before discharge and is known to contain microplastics. For instance, numerous industrial parks along the Hong River—as well as craft villages involved in activities such as textile manufacturing and plastic recycling—are considered significant sources of microplastics (footnote 60). A primary source of microplastic pollution in the Citarum River is the textile industry, where textile companies either do not treat wastewater properly, or dispose of it directly into the river. Crucially, even if producers have wastewater treatment systems, these are unable to completely remove microplastics, exacerbating the issue.[61]

Agricultural operations have also been found to be a contributor to microplastics, with high levels of microplastics detected in waters surrounding significant agricultural areas (footnote 58). Microplastics are also associated with the weathering of fishing lines and nets, such as the fragmentation of freshwater fish cage nets along the Citarum River (footnote 61).

---

[58] UNEP. 2023. *Plastic Pollution.*
[59] PlasticBank. 2023. *Which Country Is the Most Accountable for Ocean Plastic?*; K. Afkarina et al. 2020. Distribution and Environmental Risk of Microplastics Pollution in Freshwater of Citarum Watershed. *E3S Web of Conferences.* 211. p. 03012. EDP Sciences.
[60] UNEP. 2023. *Plastic Pollution*; N. Le et al. 2023. *Microplastics in the Surface Sediment of the Main Red River Estuary.* Vietnam Journal of Earth Sciences, Volume 45(1), pp. 19-32.
[61] UNEP. 2023. *Plastic Treaty progress puts spotlight on circular economy.*

2.4.2   Impact of Microplastic Pollution

Microplastic toxicity arises from the toxic nature of plastic itself, as well as a propensity for microplastics to absorb harmful chemicals in polluted waters. Specifically, plastics contain additives (e.g., bisphenol-A and phthalates) that damage the endocrine system of aquatic organisms, affecting their mobility and reproduction. In addition, they absorb toxins such as organic pollutants and heavy metals from polluted waters, compounding the harm from ingesting microplastics alone with studies documenting detrimental effects on mortality and gene expression among aquatic lifeforms.[62] As microplastics are easily ingested by aquatic organisms at the lowest levels of the food chain (e.g., zooplankton), they can bioaccumulate at higher levels (footnote 61). This negatively impacts biodiversity and ecosystems in freshwater environments (footnote 62).

Human health is also placed at risk when microplastics are ingested through drinking water and consuming contaminated fish. For instance, microplastics have been found in seafood sold for human consumption, where they are typically found in the gut of aquatic organisms such as shellfish (footnote 50). Relatedly, harmful effects on human health may materialize through the ingestion of additives in plastic and/or toxins from the environment that have been absorbed by microplastics.[63]

---

[62]   T. Anh et al. 2020. Microplastics Contamination in a High Population Density Area of the Chao Phraya River, Bangkok. *Journal of Engineering and Technological Sciences.* 52(4).

[63]   UNEP. 2023. *Plastic Treaty progress puts spotlight on circular economy*; C. Haberstroh et al. 2021. Plastic Transport in a Complex Confluence of the Mekong River in Cambodia. *Environmental Research Letters.* 16(9). p. 095009.

3 THE IMPACT OF POLLUTION

Garbage along the coast of Baseco, Tondo, Manila, Philippines. The Manila Bay, small boats, houses, a church, and buildings loom in the background (photo by Al Benavente/ADB).

Pollution of Southeast Asia's rivers contributes to a reduction in biodiversity, accelerates global warming, and has deleterious effects on people's health and livelihoods. This chapter illustrates the mechanisms through which river pollution leads to these triple impacts and considers their wider consequences on our current and future well-being.

3.1   On Biodiversity

The Convention on Biological Diversity defines biological diversity as "the variability among living organisms from all sources including, inter alia, terrestrial, marine, and other aquatic ecosystems and the ecological complexes of which they are part; this includes diversity within species, between species and of ecosystems."[64] Pollution impacts biodiversity loss that contributes to far-reaching effects on humanity's welfare.

River pollution threatens biodiversity as it harms the functioning and survival of living organisms

Organic pollution reduces dissolved oxygen levels essential for sustaining aquatic life. It occurs when excess biodegradable material is introduced and oxygen is consumed by bacteria and other microorganisms that decompose them. The discharge of domestic and industrial wastewater (which typically contain large quantities of organic material)—as well as nutrients from agricultural run-offs that encourage algae growth—are key sources of organic pollution posing a threat to aquatic biodiversity.[65] Heavy metal pollution is also of concern because of their high toxicity to living organisms even at relatively low concentrations.[66]

Studies reveal losses and threats to biodiversity within major rivers of Southeast Asia

In the Chao Phraya, water quality has been degraded by pollution to the extent that only an estimated 30 of 190 indigenous fish species can reproduce in the river mainstream.[67] Further afield in the Citarum, heavy pollution has been associated with the loss of nearly 60% of the river's species since 2008.[68] In addition, threats to biodiversity have been identified within the 3S (Sesan, Srepok, and Sekong) rivers in the Lower Mekong Basin, the Tonle Sap Lake system, and the Mekong Delta, as dissolved oxygen levels in these areas have been in decline since 2000, and are at "polluted" and "very polluted" levels (footnote 43). Likewise, the rapid increase in heavy metal pollution in the Red River Delta does not bode well for the area's biodiversity (footnote 55).

Consequences of biodiversity loss are dire and far-reaching on the welfare of humanity

Biodiversity provides humanity with the fundamentals of life, from the food, water, and air required for sustenance. Biodiversity loss reduces the effective functioning of ecosystems, also affecting their services which are the numerous, life-sustaining benefits that humans derive from thriving ecosystems. For instance, well-maintained biodiversity protects food production from threats such as pests, disease, and irregular or inadequate water supply, thus promoting food security. Beyond securing basic sustenance, it also contributes to livelihoods—including tourism and agriculture—thus enhancing overall quality of life.[69] Biodiversity protection is fundamental to achieving food security, poverty reduction, and more inclusive and equitable development.

[64]   The Convention on Biological Diversity. 2006. *Article 2 Use of Terms.*
[65]   UNEP. 2016. *A Snapshot of the World's Water Quality—Towards a Global Assessment.*
[66]   N. Nguyen and I. Volkova. 2018. Assessment of Heavy Metal Pollution in Water and Sediments in the Red River Delta (Vietnam). *IOP Conference Series: Materials Science and Engineering.* 451(1). p. 012203. IOP Publishing.
[67]   D. Allen et al. 2018. IUCN.
[68]   *The Diplomat.* 2018. Indonesia's Citarum: The World's Most Polluted River.
[69]   *South China Morning Post.* 2022. Why Everyone Should Care about Biodiversity Loss.

The socioeconomic case for more ambitious biodiversity action is clear. Ecosystem services are estimated to provide benefits of $125 trillion–$140 trillion per year, representing more than 1.5 times the size of the global gross domestic product (GDP). The costs of inaction on biodiversity loss are high and are anticipated to increase. Estimates reveal that the world lost $4 trillion–$20 trillion per year in ecosystem services from 1997 to 2011 owing to land-cover change, and an estimated $6 trillion–$11 trillion per year from land degradation. Specifically, biodiversity loss can lead to reduced crop yields and fish catches, increased economic losses from flooding and other disasters, and the loss of potential new sources of medicine. Conserving, sustainably using, and restoring biodiversity are vital to achieving many other policy objectives, including human health, climate change mitigation and adaptation, disaster risk reduction, as well as water and food security.

The economic values associated with biodiversity can be considerable: for example, the annual market value of crops dependent on animal pollination is $235 billion–$577 billion.[70] As the wide-ranging ecosystem services that biodiversity provides are at risk of being impaired through river pollution, there is a need for increased awareness and commitment to protect and enhance the biodiversity of river environments.

3.2 On Climate Change

The impact of pollution contributes to the wide-ranging and damaging effects of global warming.

River pollution catalyzes the emission of greenhouse gases

Emissions of greenhouse gases (i.e., carbon dioxide, methane, and nitrous oxide) from rivers have been found to increase when rivers are polluted. In their natural state, rivers produce carbon dioxide and methane through the bacterial decomposition of organic matter, while nitrous oxide is released by nitrifying and denitrifying organisms.[71] However, the discharge of wastewater and nitrogen-based fertilizers provide additional inputs for these processes, thus elevating the amount of greenhouse gas emissions from rivers.[72]

Furthermore, pollution-induced emissions are significant. A study of an urban river system in Ecuador revealed that the combined global warming potential of greenhouse gas emissions was 10 times higher when water quality was very heavily polluted, as opposed to being acceptable (footnote 65). The unabated pollution of rivers in Southeast Asia thus has significant potential to exacerbate global warming.

The impact of global warming has severe welfare consequences

River pollution also contributes to global warming, which has wide-ranging and severe effects. An increase in temperatures alters weather patterns that affect ecosystems supporting life on Earth and encourages extreme weather events. Its consequences are being realized through disasters (e.g., droughts, water scarcity, wildfires) and risks (rising sea levels, declining biodiversity), which do not bode well for the future. From an economic perspective, countries in Southeast Asia are expected to experience slower growth because of global warming, with Malaysia and Thailand projected to have growth of 20% below a no-warming scenario by 2050.[73]

---

[70] OECD. 2019. *Biodiversity: Finance and the Economic and Business Case for Action.*

[71] L. Ho et al. 2020. Effects of Land Use and Water Quality on Green House Gas Emissions from an Urban River System. *Biogeosciences Discussions.* pp. 1–22.

[72] UNEP. 2016. *A Snapshot of the World's Water Quality—Towards a Global Assessment*; R. Smith et al. 2017. Influence of Infrastructure on Water Quality and Green House Gas Dynamics in Urban Streams. *Biogeosciences.* 14(11). pp. 2831–2849; X. Hao et al. 2021. Greenhouse Gas Emissions from the Water-Air Interface of a Grassland River —A Case Study of the Xilin River. *Scientific Reports.* 11(1). p. 2659.

[73] United Nations Framework Convention on Climate Change (UNFCCC). 2022. *What Is the Triple Planetary Crisis.*

3.3 On Livelihood and Health

The deleterious effects of pollution on the livelihoods and health of people reduce the quality of life.

Within the major rivers of Southeast Asia, the livelihoods of communities reliant on agriculture have been impacted by water quality degradation

In the Citarum basin, rice farmers have observed a 50% reduction from their normal harvests and a decline in rice quality, which they associate with the release of dyes from textile factories.[74] In Viet Nam, the seafood industry has been threatened by water pollution from municipal, industrial, and industry wastewater discharge, which has resulted in heavy losses from the death of fish and mollusks.[75] At the economy-wide level, a study by the World Bank estimates that water pollution could cost Viet Nam up to 3.5% of its annual GDP by 2035.[76] Within the Chao Phraya basin, aquaculture farms have experienced declining productivity alongside mass mortality incidents from wastewater intrusion, with some farmers reporting productivity and income losses from 30% to 90%.[77]

Apart from impacting livelihoods, river pollution has placed community health at risk, with an increased danger of contracting fecal–oral diseases and cancer

Communities are exposed to an elevated risk of contracting fecal–oral diseases (e.g., cholera, typhoid, diarrheal diseases) because of the discharge of excreta. For instance, fecal indicator bacteria levels in the Red River Delta have been found to exceed World Health Organization guidelines that relate to a 10% risk of contracting gastrointestinal illness after a single exposure (footnote 33). The 9 million people who live in close contact with the Citarum River are similarly affected by high fecal indicator bacteria levels that are 5,000 times above mandatory limits, with many suffering from dermatitis, contact rashes, and intestinal problems.[78] The attendant financial burden on households is sizable, as the introduction of on-site sanitation facilities to reduce the discharge of human excreta into the Citarum has been estimated to reduce healthcare expenditures by 36%.[79] Communities in the Mekong Delta are likewise exposed to similar risk, as coliform contamination (an indicator of pathogens causing fecal–oral diseases) is a serious long-standing problem, with local area studies revealing contamination levels up to 11 times higher than Vietnamese water quality standards.[80]

Communities also face a heightened risk of contracting cancer from heavy metal pollution. Communities in the Chao Phraya basin have been assessed to be at high risk through the direct and indirect ingestion of and dermal contact with these pollutants.[81] In Viet Nam, residents of the Red River's estuary are exposed to such carcinogenic risk with heavy metal pollution resulting in an estimated 53 additional cases of malignant growth per 10,000 people. Within the estuary, Ha Noi's residents are exposed to the highest degree of risk, with lead pollution from the sewage of metallurgical enterprises being a key risk factor.[82]

---

[74] Cita-Citarum. 2012. *Indonesian Lives Risked on "World's Most Polluted" River.*

[75] Vietfish Magazine. 2016. *Seafood Industry in Vietnam Status and Solution for Aquatic Environment.*

[76] *Vietnam News. 2019.* Vietnam Looks towards a Safe, Clean and Resilient Water System.

[77] D. Marks et al. 2022. Increasing Livelihood Vulnerabilities to Coastal Erosion and Wastewater Intrusion—The Political Ecology of Thai Aquaculture in Peri-Urban Bangkok. *Geographical Research.* 61(2). pp. 259–272.

[78] *The Guardian. 2020.* Rotten River: Life on One of the World's Most Polluted Waterways—Photo Essay.

[79] ADB. 2013. *Downstream Impacts of Water Pollution in the Upper Citarum River, West Java, Indonesia.*

[80] T. Nguyen et al. 2022. Major Concerns of Surface Water Quality in South-West Coastal Regions of Vietnamese Mekong Delta. *Sustainable Environment Research.* 32(1). pp. 1–14; N. Giao and V. Q. Minh. 2021. Evaluating Surface Water Quality and Water Monitoring Variables in Tien River, Vietnamese Mekong Delta. *Jurnal Teknologi.* 83(3). pp. 29–36.

[81] N. Ariyakanon. 2021. Correlation of Water Quality and Heavy Metal Concentration in Chao Phraya River Basin—An Effect on Human Health. *International Journal of Science and Innovative Technology (IJSIT).* 4(2).

[82] N. Nguyen. 2018. Risks Assessment of Water Pollution at Estuary Area of Red River (Vietnam). *IOP Conference Series: Materials Science and Engineering.* 451(1). p. 012204. IOP Publishing.

4 ADDRESSING POLLUTION: GLOBAL AND REGIONAL INSIGHTS

Central Mekong Delta Region Connectivity Project. Cao Lanh Bridge is a cable-stayed bridge having a length of 2 kilometers spanning the Tien River connecting Cao Lanh City and Lap Vo District, Dong Thap Province (photo by Viet Tuan/ADB).

Preventing water pollution and conserving water are extremely important to ensure a continuing abundance of water that is safe to use for current and future generations. This chapter focuses on summarizing insights from global best practices as well as regional circumstances and challenges in combating the problem of river pollution in Southeast Asia. Four key categories of approaches are highlighted: (i) nature-based solutions, (ii) policy measures, (iii) institutional arrangements, and (iv) financing mechanisms (Table 5).[83]

Table 5: Summary of Approaches to Combat River Pollution in Southeast Asia

	Approach	Description
Nature-Based Solutions		
1	Constructed wetland	The construction of treatment systems that use natural processes involving wetland vegetation, soils, and their associated microbial assemblages to improve water quality.
2	Mangrove reforestation	The regeneration of mangrove forest ecosystems in areas where they have previously existed.
3	Others	Nature-based solutions encompass a wide range of approaches that leverage natural processes and ecosystems to address environmental and sustainability challenges. For example, riparian buffers, green roofs, permeable pavements, vegetated swales, algal turf scrubbers, "Room for the River" (managing and restoring the natural flow and course of rivers to reduce the risk of flooding and enhance the ecological health of river ecosystems).
Policy Measures		
3	Payment for ecosystem services	This occurs when the beneficiaries or users of an ecosystem service make payments to the providers of that service.
4	Water quality credit trading	A market-based approach that allows entities to buy and sell pollution reduction credits, encouraging those who can reduce pollution in rivers more efficiently to do so.
5	Water pollution taxes	Taxes are typically imposed on entities that discharge pollutants into water bodies such as rivers, lakes, and oceans.
6	Extended producer responsibility	The concept is that manufacturers and producers of goods are responsible for managing the environmental impact of their products throughout their entire lifecycle.
Institutional Arrangements		
7	Intergovernmental cooperation	The collaborative efforts and agreements between government entities across national borders to address shared issues related to the pollution and protection of rivers.
8	Intragovernmental cooperation	The coordination among various departments, agencies, or divisions within a single government to develop and implement policies, regulations, and strategies to combat river pollution and ensure effective governance.
Financing Mechanisms		
9	Blue bonds	A debt instrument that is issued to support investments in healthy oceans and blue economies.
10	Public–private partnership	A long-term agreement between public and private entities that allows the private sector to provide public services.
11	Blended finance	The strategic combination of public and private sector funds, investments, or resources to support and finance projects, initiatives, or solutions aimed at mitigating or addressing river pollution.

Source: Literature review, EY analysis.

---

[83] These potential approaches have been selected and developed based on an extensive review of professional and academic literature, along with expert interviews that reinforced the findings.

4.1 Nature-Based Solutions

The United Nations Environment Assembly defines nature-based solutions as actions to protect, conserve, restore, sustainably use and manage natural or modified terrestrial, freshwater, coastal and marine ecosystems which address social, economic and environmental challenges effectively and adaptively, while simultaneously providing human well-being, ecosystem services, resilience and biodiversity benefits.[84] The benefits of implementing nature-based solutions (NBS) extend beyond water quality and into water quantity improvement, biodiversity protection, and climate change resilience, NBS can also provide co-benefits such as biodiversity conservation, recreational opportunity, and job creation. These are value-added that cannot be achieved with conventional engineering-based solutions solely. These are typically adopted in conjunction with conventional water infrastructure to bring about more sustainable outcomes. While NBS has its advantage over grey infrastructure, it is not without its drawbacks, which include the unpredictability of system dynamics, the need for considerable coordination for effective implementation, long duration before positive results are observed, and reduced effectiveness when pollution loads are heavy.[85]

There are three main ways in which NBS can be harnessed:

(i) using and protecting natural ecosystems,

(ii) restoring ecosystems, and

(iii) creating new ecosystems.

This study highlights the implementation of constructed wetlands and mangrove reforestation to combat river pollution given that they have been utilized effectively globally to address water pollution.

4.1.1 Constructed Wetlands

Overview

Wetlands are areas where water either covers the soil's surface year-round or is intermittently present, including during the growing season.[86] They are recognized as some of the most productive ecosystems globally, often comparable to the biodiversity of rainforests and coral reefs.[87] Wetlands can manifest in various forms including rivers, marshes, bogs, mangroves, mudflats, ponds, swamps, billabongs, lagoons, lakes, and floodplains. It is important to note that extensive wetland areas often encompass a combination of diverse freshwater systems.[88]

While wetlands can occur naturally in the wild, they can also be artificially constructed. Constructed wetlands have been adopted as an innovative method for sustainable wastewater treatment in more than 50 countries,[89] including Germany, Indonesia, Thailand, the United Kingdom (UK), and the United States (US).[90] Constructed wetlands are known for their simplicity, low cost relative to gray infrastructure, and proven effectiveness as a reliable wastewater treatment method.[91]

---

[84] United Nations Environment Assembly of the United Nations Environment Programme. 2022. Resolution adopted by the United Nations Environment Assembly on 2 March 2022.

[85] CSEnvironment-DHI et al. 2018. *Nature-Based Solutions for Water Management: A Primer.*

[86] United States Environmental Protection Agency (EPA). *What Is a Wetland?*

[87] EPA. *How Do Wetlands Function and Why Are They Valuable?*

[88] World Wide Fund for Nature (WWF). *What Is a Wetland? And 8 Other Wetland Facts.*

[89] H. Wu et al. 2023. Constructed Wetlands for Pollution Control. *Nature Reviews Earth and Environment.* 4. pp. 218–234.

[90] J. Vymazal. 2022. The Historical Development of Constructed Wetlands for Wastewater Treatment. *Land.* 11(2). pp. 174.

[91] EKBY. *Constructed Wetlands.*

Poor wastewater management practices and runoff from fertilizer and application in agricultural regions have led to elevated nitrate levels in rivers. These heightened nitrate levels pose a threat to drinking water safety and contribute to issues such as algal blooms and the degradation of aquatic ecosystems. Wetlands can mitigate these adverse effects by converting harmful nitrates in the water into harmless nitrogen gas, which is subsequently released into the atmosphere. The microorganisms in the sediment and vegetation within the soil are adept at breaking down various types of waste, eradicating pathogens, and reducing nutrient and pollution levels in the water. This process improves water quality and allows cleaner water to flow downstream.[92]

There are mainly three configurations of constructed wetlands (Figure 2):[93]

(i) Horizontal Flow Constructed Wetlands

(ii) Vertical Flow Constructed Wetlands

(iii) Hybrid Flow Constructed Wetlands

Figure 2: Three Main Configurations of Constructed Wetlands

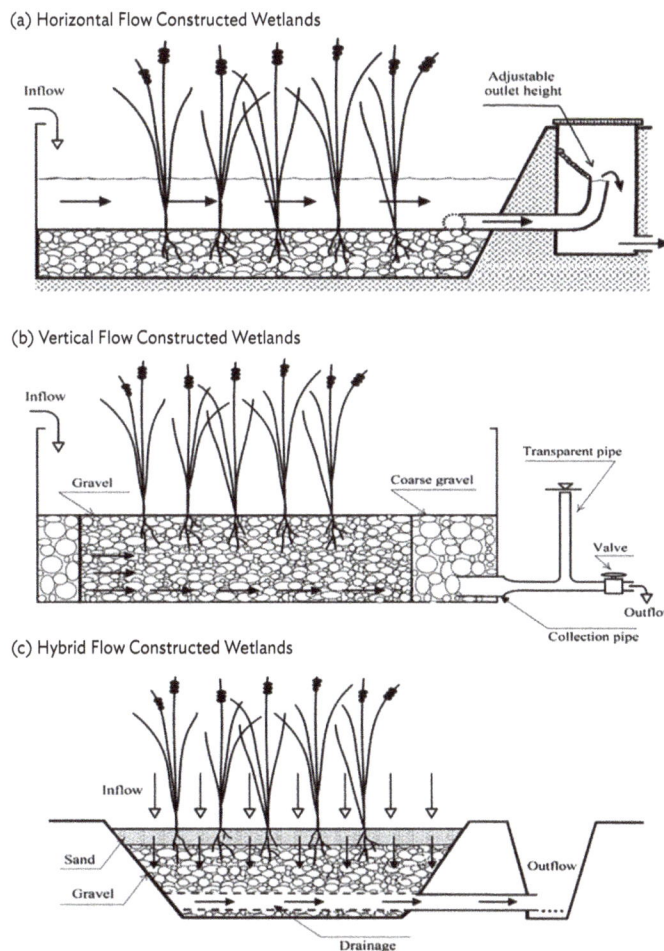

(a) Horizontal Flow Constructed Wetlands

(b) Vertical Flow Constructed Wetlands

(c) Hybrid Flow Constructed Wetlands

Source: H. Affum. 2022. *New CRP: Development of Radiometric Methods for the Measurement of Constructed Wetlands Hydrodynamics (F22076)*. International Atomic Energy Agency (IAEA).

---

92 Flood. 2020. *Restoring wetlands near farms would dramatically reduce water pollution*. UIC Today.
93 UN–Habitat. 2008. *Constructed Wetlands Manual*.

Horizontal flow constructed wetlands are effective for pollutant removal but require more space. Vertical flow constructed wetlands are space-efficient and highly effective at pollutant removal, but more complex to construct. Hybrid flow constructed wetlands combine elements of both horizontal and vertical flow, offering a good balance.[94] Further variations in the configurations of wetlands are shown in Figure 3.

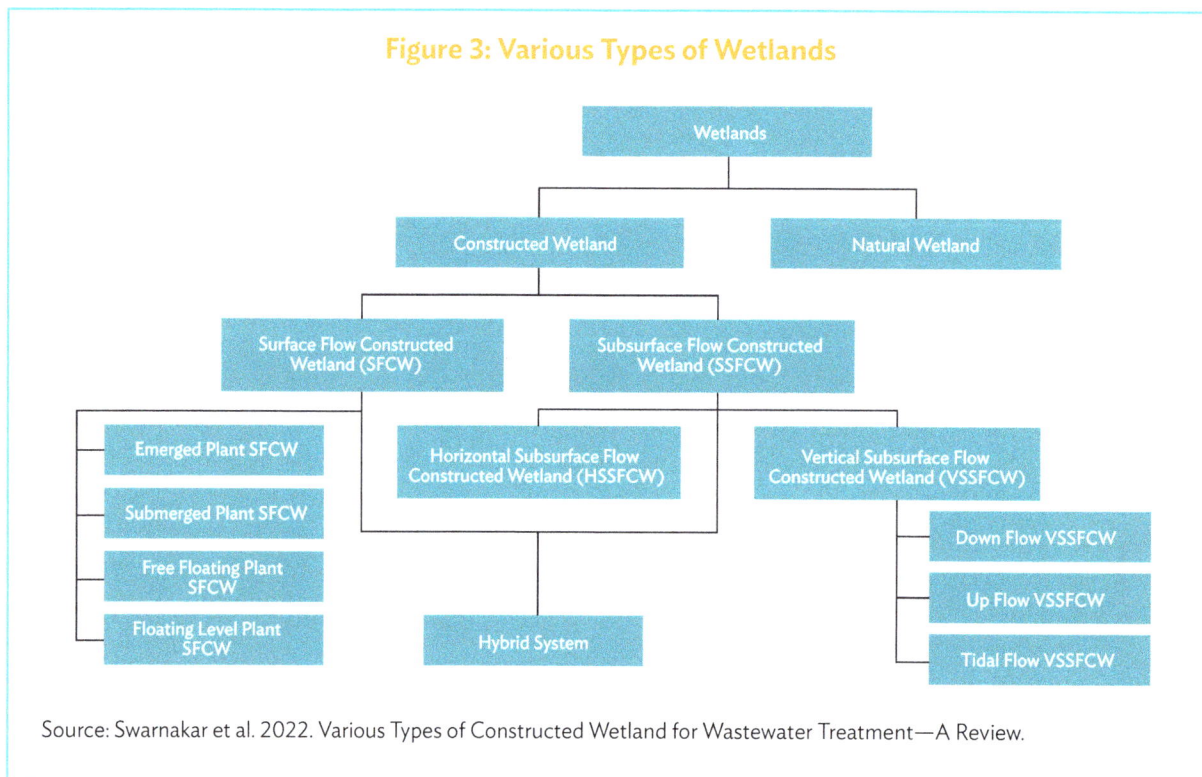

Figure 3: Various Types of Wetlands

Source: Swarnakar et al. 2022. Various Types of Constructed Wetland for Wastewater Treatment—A Review.

Background

Early research into wetland viability for wastewater treatment started in Germany in the 1950s and spread to Hungary, the Netherlands, and the US in the 1960s and 1970s. The rapid global expansion of constructed wetlands for wastewater treatment occurred during the 1980s and 1990s. Throughout the 1990s, numerous international conferences that focused on this technology were held throughout Europe, Asia (India, Nepal, and the People's Republic of China [PRC]), Australia, and North and South America. During the last decade of the 20th century, constructed wetland technology extended to all continents using all types of constructed wetlands (surface flow, horizontal subsurface flow, and vertical flow).[95]

Constructed wetlands are recognized as a certified wastewater treatment method in many countries. Many have released their guidelines for constructed wetlands, such as Germany,[96] New Zealand,[97] the UK,[98] the US,[99]

---

[94]   Swarnakar et al. 2022. Various Types of Constructed Wetland for Wastewater Treatment—A Review.

[95]   J. Vymazal. 2022. The Historical Development of Constructed Wetlands for Wastewater Treatment. *Land.* 11(2). pp. 174.

[96]   J. Nivala et al. 2018. The New German Standard on Constructed Wetland Systems for Treatment of Domestic and Municipal Wastewater. *Water Science and Technology.* 78(11). pp. 2414–2426.

[97]   C. Tanner. *Constructed Wetland Guidelines.* NIWA.

[98]   J. Ellis et al. 2003. *Guidance Manual for Constructed Wetlands.* Environment Agency.

[99]   EPA. 2015. *A Handbook of Constructed Wetlands.*

and even intergovernmental organizations like the UN.[100] Constructed wetlands are also gaining attention for sustainable water management in urban areas, aligning with the circular economy and the concept of "sponge" cities.

As of 2023, there are 43 cities in 17 countries globally accredited as Wetland Cities, highlighting the increasing understanding of the benefits that wetlands provide.[101] In 2022, Surabaya in Indonesia achieved accreditation as a Wetland City, primarily attributed to its extensive wetland ecosystems covering an area of 1,722 km². These wetlands provide vital support to a diverse range of wildlife. Surabaya has also initiated urban planning efforts in collaboration with community associations. These initiatives aim to tackle a variety of challenges including the mitigation of river and coastal pollution, addressing seasonal water scarcity, and managing urban flooding (Figure 4).[102]

Figure 4: Example of a Constructed Wetland in Yanweizhou Park in Jinhua, People's Republic of China

Source: L. Garfield. 2017. China is building 30 "sponge cities" that aim to soak up floodwater and prevent disaster. *Business Insider*. 10 November.

Implementation

Constructed wetlands can be established by any entity—local or foreign stakeholder—invested in the well-being of the local population expected to benefit from the presence of a wetland. Typically, these entities finance local contractors, engineers, and scientific experts to develop the wetland. Local governmental agencies typically spearhead the wetland construction efforts, with financing support from both international and private donors.[103] Some examples in the US include the following:

---

[100]   UN-HABITAT. 2018. *Constructed Wetlands Manual*. UN-HABITAT Water for Asian Cities Programme Nepal, Kathmandu.
[101]   Ramsar. *Wetland City Accreditation*.
[102]   Cities with Nature. *Celebrating the Importance of Wetland Cities on World Wetlands Day*.
[103]   EPA. 1993. *Constructed Wetlands for Wastewater Treatment and Wildlife Habitat: 17 Case Studies*.

- **Carolina Bays:** The US Environmental Protection Agency and South Carolina Department of Health and Environmental Control awarded grants for the planning, pilot testing, design, and construction of the full-scale Carolina Bay Natural Land Treatment Program; today, the wetlands are maintained by a regional water utility company, the Grand Strand Water and Sewer Authority.[104]

- **Houghton Lake, Michigan:** The US National Science Foundation and Houghton Lake Sewer Authority sponsored the Wetland Ecosystem Research Group at the University of Michigan to design, build, and operate a full-scale wetlands system; today, the wetlands are maintained by the Houghton Lake Sewer Authority.[105]

Applicability to Southeast Asia

Southeast Asia is home to more than 54% of the world's peatlands—a specific type of wetland—and this prominence is because of the suitability of its climate.[106] The use of wetlands in Southeast Asia to combat river pollution is still in the exploratory phase but is garnering momentum. Some state-run wetland restoration programs exist, but these initiatives often serve purposes beyond addressing river pollution, such as carbon sequestration, flood prevention, and fire prevention. One notable example is Indonesia's nationwide peatland restoration initiative. Indonesia is one of the leading nations in the region in the restoration of wetlands. As a result of widespread peatland fires in 2015, Indonesia established a Peatland Restoration Agency in 2016, tasked with restoring 2.6 million hectares of degraded peatlands.[107]

Several studies highlight the strong potential of constructed wetlands for Southeast Asia.[108] For example, an assessment of three full-scale constructed wetland systems across Thailand (Koh Phi Phi, a world-renowned international tourist and holiday resort; Sakon Nakhon, a northeastern provincial capital; and Ban Pru Teaw, a small post-tsunami village on the Andaman Coast) managed to provide nitrogen removal rates of 86% and biological demand reduction by 87% (Figure 5).[109]

Challenges and key success factors

Three key challenges that surface are the following:

(i) **Biological complexity.** Wetlands exhibit significant biological complexity, primarily because there is a limited understanding of the natural functions of wetlands and how native plants and animals respond to wastewater, thus making it difficult to predict treatment performance, manage ecological interactions, and maintain desired treatment outcomes.[110] Additionally, their designs can be complex given their expansive land coverage that creates a need to justify them against competing land uses (footnote 91).

[104] NSCEP. 2004. *Carolina Bays: A Natural Wastewater Treatment Program.* EPA.

[105] NSCEP. 1993. *Natural Wetlands for Wastewater Polishing: Houghton Lake, Michigan.* EPA.

[106] D. Sulaeman et al. 2022. *Mapping, Management, and Mitigation: How Peatlands Can Advance Climate Action in Southeast Asia.* WRI.

[107] UNDP. Support Facility for the Peat Restoration (Badan Restorasi Gambut).

[108] P. Hamel and L. Tan. 2022. Blue–Green Infrastructure for Flood and Water Quality Management in Southeast Asia: Evidence and Knowledge Gaps. *Environmental Management.* 69. pp. 699–718.

[109] K. Møller et al. 2012. Economic, Environmental and Socio-cultural Sustainability of Three Constructed Wetlands in Thailand. *Environment and Urbanization.* 24(1). pp. 305–323.

[110] B. Gopal. 1999. Natural and Constructed Wetlands for Wastewater Treatment: Potentials and Problems. *Water, Science and Technology.* 40(3). pp. 27–25.

(ii)   **Lack of sustainable funding**. Adequate planning and monitoring are critical for the long-term success of wetlands in wastewater treatment, particularly because of the lack of predictability in treatment efficiency. However, this can prove complicated because of a lack of funding.[111] Many constructed wetlands around the world are managed by local governments or agencies responsible for watershed management. Many local governments lack the financial resources to undertake large-scale wetland restoration projects without external funding options.

(iii) **Insufficient political will and coordination**. The implementation of a constructed wetland requires the involvement of many stakeholders such as government, municipal offices, and local communities. Each stakeholder needs to be able to recognize the benefits of constructed wetlands on the local community,[112] especially given that the benefits are not as immediate compared with gray infrastructure.[113] As a result of the benefits being delayed, communicating benefits presents significant difficulties, hindering the implementation of constructed wetlands.

Figure 5: The Flower and the Butterfly Constructed Wetland System at Koh Phi Phi, Thailand

Source: H. Brix et al. 2011. The flower and the butterfly constructed wetland system at Koh Phi Phi—System design and lessons learned during implementation and operation. *Ecological Engineering*, Volume 37(5), pp. 729-735.

---

[111]   A. Canning et al. 2021. Financial Incentives for Large-Scale Wetland Restoration: Beyond Markets to Common Asset Trusts. One Earth. 4(7). pp. 937–950.

[112]   T. Musasa et al. 2023. The Role of Stakeholder Participation in Wetland Conservation in Urban Areas: A Case of Monavale Vlei, Harare. *Scientific African*. 19.

[113]   OECD. 2020. Nature-Based Solutions for Adapting to Water-Related Climate Risks. *OECD Environment Policy Paper, Number 21*.

Considering the challenges in developing constructed wetlands, three key success factors can ensure the long-term success of such projects or initiatives:

(i) **Thorough planning.** Every site is unique, requiring a site-specific design for a constructed wetland system oriented toward creating a biologically and hydrologically functional system. Plans should also include detailed instructions for implementing a contingency plan if the system does not achieve its expected performance within a specified time (footnote 99).

(ii) **Effective management and maintenance.** Regular maintenance and management of the wetland are vital for long-term success. This includes monitoring water quality, vegetation health, and wildlife populations, as well as addressing any issues like invasive species or clogging. Proper management—supported by adequate funding—ensures the wetland continues to function as intended, and its benefits can be realized for an extended period.[114]

(iii) **Community engagement.** Involving and gaining support from local communities and stakeholders is essential. Community engagement helps raise awareness of the wetland's benefits, fosters a sense of ownership, and encourages responsible behavior around the wetland. It can also provide resources and volunteers for maintenance and monitoring efforts. Effective community involvement can reduce the need for constant funding, especially when community members understand the benefits they receive. Moreover, it is likely to increase the probability of action being taken for the community.[115]

Box 2 presents the case of a constructed wetland system in southern Thailand.

Box 2: Constructed Wetland System at Koh Phi Phi, Thailand

Background of the project

In 2007, a constructed wetland system was implemented on the tourist island of Koh Phi Phi in southern Thailand. The system was a project of significance showing the potential for aesthetical integration of constructed wetland systems in the built environment. The system comprised a wastewater collection system for the business and hotel area of the island, a pumping station and a pressure pipe system to the treatment facility, a multi-stage constructed wetland system, and a system for the reuse of treated wastewater. The treatment chains consisted of vertical flow, horizontal subsurface flow, and free water surface flow units. Because the treatment system was located at the center of the island, surrounded by resorts, restaurants, and shops, the project had systems designed with terrains as scenic landscaping. Wastewater was treated to meet Thailand's effluent standards for total suspended solids and nitrogen, but the project did not meet the standards for oil and grease and biochemical oxygen demand (BOD) because of inadequate pre-treatment and prior removal of oil and grease. Several challenges during construction and operation caused problems with clogging of the vertical-flow-constructed wetlands and caused odor problems. The project prepared safeguards but did not effectively activate them, mainly because no key person or key authority took responsibility for managing the system.

continued on next page

---

114 A. Gorgoglione and V. Torretta. 2018. Sustainable Management and Successful Application of Constructed Wetlands: A Critical Review. *Sustainability*. 10(11). p. 3910.

115 Government of Australia, Department of the Environment. 2016. *Wetlands and the community*. DCCEEW.

Box 2 *continued*

In terms of implementing the project, it had been anticipated that close follow-up and adjustments would be required, especially as not all stakeholders volunteered to have oil and grease traps installed on their private land during the construction phase. The project established several post-construction safeguards to assist the municipality in operating and maintaining the system and to reduce the risk of system failure. Safeguards included the following:

(i) A performance bond corresponding to 10% of the construction cost that enabled the municipality and the donor—the Danish International Development Agency—to require the contractor to rectify mistakes or omissions discovered within the first year after construction.

(ii) A 3-year post-supervision contract was signed with a local wastewater management expert to closely monitor and supervise the technical operation and maintenance tasks that would arise and inform the municipality and donor if immediate action was to be enacted to sustain the system.

(iii) A 5-year budget covering operation and maintenance tasks was provided to the municipality with bi-yearly installments of B250,000 ($6,300) to provide time for the municipality to include the full costs in its normal budget and to make it possible for the donor to withhold budgets if necessary.

(iv) The implementation of public relations activities informing about the importance of oil and grease traps and the collection system, as well as a community-based committee responsible for the operational and financial aspects of the wastewater management system.

Challenges

Although the treatment efficiency was observed to be quite good on all measured indicators, the project suffered from several operational problems. In the early years of operation of the wastewater management facilities, a series of operation and maintenance issues originating from the construction works manifested. The project did not satisfactorily resolve the problems despite efforts by operators to remediate the system.

(i) While it seemed as though the residents and business owners were generally satisfied with the system, the community-based committee, the municipality, and the mayor remained passive and did not actively take responsibility for the management and problem-solving.

(ii) No public relations support was provided to increase awareness of the importance of the collection system and the oil and grease traps.

(iii) The donor did not effectively activate the safeguards but seemed more focused on placing responsibility for the malfunction of the system. The performance bond safeguard was not activated within a year after commencement to make the contractor rectify the reported issues.

(iv) It appeared that the municipality did not take charge of the treatment system and that there was no key person to resolve issues associated with the system.

Overall, the lack of action was deemed to have stemmed from a complex mix of reasons. Although the project had financial and technical tools in place, it wasted significant time and energy on reporting and carrying out discussions without actual execution to solve the observed and manageable operation issues. This resulted in an escalation of odor problems associated with the treatment system, and the challenge of rectifying the system grew bigger day by day.

continued on next page

Box 2 *continued*

Key lessons learned

Constructed wetlands have a large potential in developed and developing countries as robust, reliant, and cost-effective wastewater treatment systems mutually addressing programmatic synergies with the social, spatial, and environmental dimensions of urban development. From a technical point of view, the appropriateness of such systems has been thoroughly documented during the last 30 years. Still, a rigorous desktop design and an appropriate set of safeguards do not eliminate the risk of system failure.

The risk of mistakes during construction works, the absence of adequate regulations and people awareness, the local history of past successes and failures, knowledge, practices, and skills are factors influencing the chances for long-term sustainability. In a complex social setting with many stakeholders, motives, and interests—as the one experienced at Koh Phi Phi—the continuous commitment and involvement of stakeholders, and the shared and individual responsibility for the successful performance of the system are of paramount importance. At Koh Phi Phi, the lack of a key person or a key authority taking responsibility for managing the system has been regarded as the most important reason behind the system not being as successful as intended 3 years after implementation. This circumstance points to the social and institutional dimension as the next crucial step to overcome in the task of developing a more sustainable wastewater management practice in developing countries.

B = Thai baht.
Sources: K. Møller et al. 2012. Economic, Environmental and Socio-cultural Sustainability of Three Constructed Wetlands in Thailand. *Environment and Urbanization.* 24(1). pp. 305–323; Hans Brix et al. 2010. *The flower and the butterfly constructed wetland system at Koh Phi Phi—System design and lessons learned during implementation and operation.*

4.1.2 Mangrove Reforestation

Overview

Mangroves are shrub and tree species that live along shores, rivers, and estuaries in the tropics and subtropics.[116] Many mangrove forests can be recognized by their dense tangle of prop roots that make the trees appear to be standing on stilts above the water.[117]

Mangroves help to improve water quality in the following ways:[118]

(i)   **Nutrient uptake**. Mangroves possess complex root systems that function as natural filters, effectively removing harmful substances like nitrates, phosphates, and other pollutants from the water. This purification process significantly enhances the quality of water flowing from rivers and streams into estuaries and oceans.

---

[116]   American Museum of Natural History. *What's a Mangrove? And How Does It Work?*
[117]   Government of the United States, National Ocean Service. 2023. *What is a mangrove forest?*
[118]   The Nature Conservancy. 2020. *The Importance of Mangroves.* 4 May.

(ii) **Sediment filtration**. Mangrove root systems contribute to sediment stabilization by reducing erosion and encouraging sediment deposition. Their above ground roots slow down water flows and promote sediment retention, preventing runoff into adjacent water bodies.

(iii) **Erosion buffers**. Mangroves serve as natural infrastructure and act as protective buffers during extreme weather events, such as monsoon storms and tsunamis. By absorbing the impact of storm surges and wave energy, they play a critical role in preventing the displacement of sediments and pollutants.

The benefits of mangroves extend beyond water quality to include carbon sequestration and wildlife habitats. Mangroves store up to five times as much organic carbon as tropical upland forests because of their high productivity and slow soil decomposition rates. This significantly enhances their capacity to capture and store organic carbon, especially in their soils.[119] Mangroves also provide a habitat for numerous species of fish, crustaceans, and other marine life, which can then serve as a food source for humans.

Background

Mangrove forests were estimated to have covered 75% of tropical coasts worldwide, but today, the global range of these forests is estimated to be less than 50% of their original total cover. These losses have largely been attributed to anthropogenic pressures such as over-harvesting for timber and fuel-wood production. Another major threat to mangrove forests is their conversion into areas for aquaculture. For instance, in the Indo-western Pacific region, 1.2 million hectares of mangroves have been converted to aquaculture ponds during 1980–2020.[120]

Rehabilitation and restoration of mangroves have been initiated in regions including Southeast Asia (e.g., Indonesia, Thailand, Viet Nam), East Asia, and South America. The rationale for rehabilitating or restoring mangroves includes the protection of ecosystem services, including the creation or maintenance of forest stands for high yields, coastal protection, landscaping, conservation of biodiversity, and compliance with laws that require it (e.g., local regulations mandating "No Net Loss" of wetlands following development projects) (Figure 6).

Some of the largest nations on the planet are also capitalizing on the benefits of mangroves to shore up coastline defenses and increase biodiversity. During the early 1950s to 2000, the PRC lost more than half of its mangroves, but as of 2023, 75% of them are in protected areas, well above the global average of 42%. Countrywide, in the first two decades of the 21st century, there has been a net increase of 4,850 hectares, a 23% jump.[121]

In the US, the natural sea defenses offered by mangroves are being boosted along the Florida coastline. Hurricanes, tropical storms, winds, and flooding cause damage costing over $730 billion globally every year, about $50 billion of which is in the US alone. Restoring mangrove forests is seen as a way of reducing those costs and protecting more than 15 million people worldwide (Figure 7).[122]

Implementation

One of the most common methods is the planting of mangrove seedlings, which can be grown in nurseries before being transplanted to the desired location. Another common method is the use of mangrove propagules, which are small pieces of a mangrove tree that can be used to grow new trees.

---

[119] M. Chatting et al. 2022. Future Mangrove Carbon Storage under Climate Change and Deforestation. *Frontiers in Marine Science*. 9.

[120] Background information was obtained from: A. Ellison et al. 2020. Mangrove Rehabilitation and Restoration as Experimental Adaptive Management. *Frontier in Marine Science*. 7. p. 327; J. Kairo et al. 2001. Restoration and Management of Mangrove Systems—A Lesson for and from the East African Region. *South African Journal of Botany*. 67(3). pp. 383–389; S. Song et al. 2023. Mangrove Reforestation Provides Greater Blue Carbon Benefit than Afforestation for Mitigating Global Climate Change. *Nature Communications*. 14. p. 756; BBC Earth. 2023. *Could Mangroves Help Save Our Planet?*

[121] BBC Earth. 2023. *Could Mangroves Help Save Our Planet?*

[122] Metropolitan Digital. *Protecting mangroves can prevent billions of dollars in global flooding damage every year.*

Figure 6: Mangrove Restoration Near the City of Semarang, Northern Java Island, Indonesia

Source: Wageningen University Research. Building with Nature along a Tropical Muddy Coast in Indonesia: Innovation in Mangrove Restoration.

Figure 7: Mangrove Replenishment Project in Florida, United States

Source: Florida Museum. Conservation: Mangroves.

Before mangrove restorations could be implemented, significant research is required to identify suitable species to be planted. For example, in Malaysia, the government commissioned the Forest Research Institute Malaysia, local public universities, and nongovernment organizations (NGOs) to conduct a research study on the status of mangroves and the status of the ecosystem before embarking on a nationwide mangrove reforestation program.[123]

Mangrove reforestation projects can be initiated by either government bodies or private entities. Government bodies often advocate for these projects to improve the livelihoods of the local population, while private entities embark on projects as part of their environmental, social, and governance initiatives or to earn carbon credits.

Examples of government-led initiatives:

- **Broward County Parks and Recreation Department**—a branch of local government—provided $5 million for the restoration of 525 hectares of mangroves at West Lake Park near Fort Lauderdale, Florida.[124]
- **The Government of the People's Republic of China** conducted a $35 million large-scale restoration of the mangrove and coral reef ecosystems in the southern coastal regions of Fujian, Guangxi, Guangdong, and Hainan.[125]

Examples of private-led initiatives:

- **OCBC Bank** has contributed 3 million Singapore dollars to the OCBC Mangrove Park, Singapore's first large-scale project to adopt the Ecological Mangrove Restoration method.[126]
- **Apple** is working in conjunction with Conservation International to preserve a 11,000-hectare mangrove forest in Colombia. This project is also the site of the first fully accounted carbon offset credit for a mangrove.[127]

Applicability to Southeast Asia

Southeast Asia has lost over 30% of its mangrove forests from 1980 to 2020, a decrease of more than 63,000 km^2. The region accounts for 42% of the world's total mangrove forests, which provide critical breeding habitats for about 75% of fish species caught in the oceans.[128]

Mangrove restoration has a long history in Southeast Asia. For example, the Matang Mangrove Forest in Peninsular Malaysia was designated in 1906 as a permanent forest reserve.[129] In the Philippines, mangrove planting dates to the 1930s for the supply of construction posts for fish weirs and fuel.[130] In Indonesia, mangrove rehabilitation started back in the 1930s for timber production.[131] In Viet Nam, direct planting of fast-growing Rhizophora apiculata (a tall-stilt mangrove) was practiced in 1978 in areas affected by the herbicide Agent Orange during the Viet Nam–US war.[132] In 2018, SK Innovation launched a "Mangrove Reforestation Project" with a local partner company—Petro Vietnam Exploration and Production—and have since planted over 1,000 saplings together.[133]

[123] H. Omar. 2020. Status of Mangroves in Malaysia. *Forest Research Institute.*
[124] Society for Ecological Restoration (SER). *USA: Florida: Mangrove Restoration at West Lake (Broward County).*
[125] SER. *China: Agenda 21 Mangrove Restoration in Southern China (Hainan, Guangxi, Fujian and Guangdong Regions).*
[126] National Parks Board. 2022. *Nparks Partners OCBC Bank in Singapore's First Large-Scale Ecological Mangrove Restoration Project at Pulau Ubin.*
[127] J. Klein. 2021. *Apple, Conservation International introduce mangrove carbon credit.* GreenBiz.
[128] K. Savi. 2020. ASEAN loses a third of mangroves in last 40 years. *The Jakarta Post.* 29 July.
[129] A. Ammar et al. 2014. Can the Matang Mangrove Forest Reserve Provide Perfect Teething Ground for a Blue Carbon Based REDD+ Pilot Project? *Journal of Tropical Forest Science.* 26(3). pp. 371–381.
[130] S. Aypa and S. Baconguis. 2000. *Philippines: Mangrove-friendly aquaculture.*
[131] V. Arifanti. 2020. Mangrove Management and Climate Change: A Review in Indonesia. *IOP Conference Series: Earth and Environmental Science.* 487.
[132] P. Hong. *Reforestation of mangroves after severe impacts of herbicides during the the Viet Nam war: the case of Can Gio.* FAO.
[133] *Skinno News.* 2023. Mangrove Reforestation in Vietnam. 11 May.

While early mangrove rehabilitation practices were focused on establishing mangrove cover for short-term economic gains (e.g., fuel and timber), today mangroves are planted for their many other benefits, such as storm surge and erosion protection, nursery habitat for many commercial fish and shellfish, and carbon sequestration.[134] A study conducted at the mangrove forest site of the King's Royally Initiated Laem Phak Bia Environmental Research and Development Project in Ban Laem District, Phetchaburi Province, Thailand, demonstrated that mangrove forests can significantly improve water quality. The project increased dissolved oxygen levels by 32% and reduced levels of phosphate (by 88%), ammonia (by 74%), and nitrate (by 64%). This underscores mangrove effectiveness as an additional natural system to enhance the efficiency of manmade wastewater treatment systems.[135]

Challenges and key success factors

The three main challenges to mangrove restoration are as follows:

(i) **Species selection**. Mangrove ecosystems are highly diverse and complex, with multiple species coexisting in a single forest. As different mangrove species exhibit varying tolerances to factors such as temperature, salinity, and elevation, it is critical to provide sufficient consideration of species selection based on site-specific environmental factors.[136]

(ii) **Plant survival.** Ensuring the survival and growth of planted mangrove seedlings is a significant challenge. Mangroves are vulnerable to threats from pests, diseases, and adverse environmental conditions. Without adequate care and monitoring, high mortality rates can occur. For example, in the Philippines, the World Bank invested $35 million to plant nearly 3 million mangrove seedlings in central Visayas during 1984–1992. However, by 1996, less than 20% of those mangroves had survived.[137]

(iii) **Long-term maintenance**. Mangrove restoration is a long-term process that necessitates continuous, regular monitoring and maintenance to evaluate the success of the restoration effort and make necessary adjustments. Mangrove ecosystems face constant threats from human activities like logging, agriculture, and urbanization, which can inflict damage on mangrove forests and complicate restoration efforts.[138]

Considering the challenges in conducting mangrove reforestation, three key success factors can ensure the long-term success of such projects or initiatives:

(i) **Species selection.** Species selection is important in mangrove reforestation because different mangrove species have different traits that can affect the success of the project. For example, some species are more tolerant of salinity, while others are more tolerant of erosion. Additionally, some species grow faster than others, and some provide more habitat for wildlife. In the Mikoko Pamoja project in Gazi Bay, Kenya, careful species selection contributed to its success by restoring mangroves that benefit both the environment and the local communities.[139]

[134] Florida Department of Environmental Protection. 2016. *Benefits of Mangroves*.
[135] O. Jitthaisong et al. 2012. Water Quality from Mangrove Forest: The King's Royally Initiated Laem Phak Bia Environmental Research and Development Project, Phetchaburi Province. *Ridern Applied Science*. 6(8). p. 1.
[136] J. Su et al. 2022. Priority Areas for Mixed-Species Mangrove Restoration: The Suitable Species in the Right Sites. *Environmental Research Letters*. 17(6).
[137] H. Waters. 2016. *Mangrove Restoration: Letting Mother Nature Do the Work*. Smithsonian Ocean.
[138] S. Emerson. 2023. *Mangrove Restoration: Challenges and Opportunities*. Dendra Systems.
[139] E. Schoof. 2020. *Case study: Community-led Mangrove Restoration and Conservation in Gazi Bay, Kenya—Lessons from Early Blue Carbon Projects (on going)*. The Commonwealth.

(ii) **Site selection.** Site selection is important in mangrove reforestation because different sites have different environmental conditions that can affect the success of the project. For example, some sites have more suitable tidal ranges and salinity levels than others. Additionally, some sites are more protected from erosion and have better soil quality.[140]

(iii) **Community engagement.** In 2010, residents of Gazi Bay, Kenya—facing a 20% loss of mangrove forests because of timber harvesting—partnered with UK charity Plan Vivo and Scotland-based Association for Coastal Ecosystem Services to launch a conservation and restoration project, which successfully protected 117 hectares of mangroves with full-time guards and engaged nearly 500 community members in planting activities. The project's success is attributed to strong community participation, transparent planning for land use, and revenue distribution, as well as efforts to address potential negative impacts on the local community (footnote 139).

Box 3 presents a mangrove restoration project in Viet Nam.

Box 3: Mangrove Restoration in Can Gio, Ho Chi Minh City, Viet Nam

Background of the project

In the 1960s, Can Gio had about 40,000 hectares of dense mangrove forests, but by 1971, they had been almost destroyed. In 1978, the City Forestry Service and the People's Committee of Can Gio District in Viet Nam's Ho Chi Minh City undertook the mangrove reforestation project with the following objectives:

(i) To restore the mangrove ecosystem previously destroyed by herbicides.

(ii) To stabilize the land and restrict erosion.

(iii) To contribute to the improvement of the environment by reducing the pollution caused by industrial waste and smoke discharge.

(iv) To create habitats for terrestrial animals and provide nursery breeding grounds for aquatic resources.

(v) To create employment and raise the income of foresters and fisherfolk through silviculture and aquaculture activities, thus improving the standard of living of local inhabitants.

(vi) To supply part of the demand for fuelwood and poles.

Impacts of the project

Rehabilitated mangroves in Can Gio serve as natural filters, effectively purifying both the air and estuary waters by trapping sediments and preventing solid waste discharge from Ho Chi Minh City into the sea. Additionally, this endeavor in mangrove rehabilitation resulted in the prevention of riverbank erosion, which led to the formation of extensive tidal sandy mud flats that support a variety of benthic habitats and facilitate shell and clam farming. At the beginning of 2000, the United Nations Educational, Scientific and Cultural Organization's World Network of Biosphere Reserves designated the Can Gio mangroves for inclusion, leading to a significant development of ecotourism in the area.

continued on next page

---

[140] M. Basyuni et al. 2018. Evaluation of Mangrove Reforestation and the Impact to Socioeconomic-Cultural of Community in Lubuk Kertang Village, North Sumatra. *IOP Conference Series: Earth and Environmental Science.* 126.

Box 3 *continued*

Mangroves Coverage in Can Gio District in 1973 and after Initial Recovery in 1989

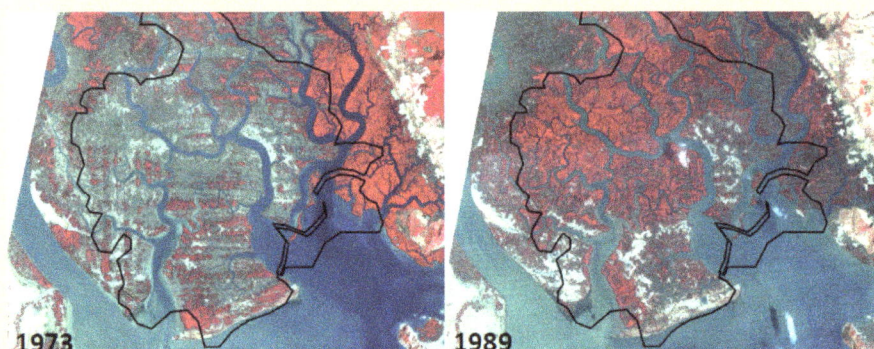

Source: V. Tuan and C. Kuenzer. 2012. *Can Gio Mangrove Biosphere Reserve Evaluation 2012: Current Status, Dynamics, and Ecosystem Services*. International Union for the Conservation of Nature (IUCN).

Challenges and key success factors

Despite the impacts, the initial project had challenges at the beginning:

(i)   The initial stages had the technical aspects of reforestation neglected, resulting in a low survival rate. Because of the lack of experience, the planting density was too high resulting in only 18,125 hectares being covered with mangroves at the beginning of 1990 despite 29,583 hectares of Rhizophora apiculata mangrove seedlings being planted in Can Gio from 1978 to 1989. However, as of 2024, the mangroves in Can Gio are more diverse in community structure than before the US–Viet Nam war because of the mixing of replanted species with naturally regenerated species.

(ii) The problem of mangrove overharvesting by the local inhabitants had to be resolved. A forestry enterprise made up of a small group of forestry workers—with insufficient means of transport and communication—was originally responsible for the protection of the vast area. In 1991, the People's Committee decided to invest funds and proper equipment to protect the forests. The forestry enterprise was converted into the management board of the City's Environmentally Protected Forests, personnel were increased, and the city put into practice a policy of land and forest allocation to households. As a result, the destruction of forests markedly decreased.

Source: P. Hong. *Reforestation of mangroves after severe impacts of herbicides during the Viet Nam war: The case of Can Gio*. Food and Agriculture Organization (FAO).

In addition to constructed wetlands and mangrove reforestation—which are typically solutions implemented near rivers—other nature-based solutions can be implemented to achieve pollution reduction and control.

(i) Green roofs. Green roofs—also known as vegetated or living roofs—have become increasingly important components of water-sensitive urban design systems. These roofs help in managing stormwater effectively, reducing urban heat, reducing energy consumption, improving air quality, and enhancing urban ecosystems (Figure 8).[141] They contribute to more sustainable and water-sensitive urban development, making cities more resilient and livable in the face of climate change and urbanization.[142] Multiple green roof projects have been implemented globally, including the Clapham Park Total Green Roof System (UK), Munich Technology Center (Germany), and Standard Chartered Bank (UK).[143]

Figure 8: Thammasat University Urban Farming Green Roof in Bangkok, Thailand

Source: Times Higher Education. 2023. *Thammasat University—Asia's largest organic rooftop farm.*

By mimicking traditional rice terraces, the Thammasat University Urban Farming Green Roof has become an all-in-one solution—as a public green space, urban organic food source, water management system, energy house, and outdoor classroom—which serves as an adaptation model for anticipated climate impacts that can be implemented and developed across the country and Southeast Asia. Any runoff is filtered through each layer of soil and later saved up in four retention ponds that can collect water up to 11,718 m³ for rooftop irrigation and future use. With each cascading level, the green roof is not only able to absorb rainwater but also slow down runoff, both for up to 20 times more than a normal concrete rooftop.[144]

[141]   Science Direct. 2023. *Green roofs as a nature-based solution for improving urban sustainability: Progress and perspectives.*
[142]   Multidisciplinary Digital Publishing Institute. 2022. *Vegetated Roofs as a Means of Sustainable Urban Development: A Scoping Review.*
[143]   International Federation of Consulting Engineers (FIDIC). 2023. *New playbook on nature-positive infrastructure highlighted at New York Climate Week.*
[144]   Times Higher Education. 2023. *Thammasat University—Asia's largest organic rooftop farm.*

(ii) Permeable pavements. Permeable pavements are a sustainable and innovative approach to urban design and construction. Permeable pavements can help manage stormwater, reducing the risk of localized flooding and erosion (Figure 9). They also allow rainwater to infiltrate the ground beneath the pavement, helping in recharging groundwater. These pavements thus contribute to water quality improvement by preventing the discharge of contaminants such as oil and heavy metals from paved surfaces into water bodies.[145] In Seoul's Songpa district, a permeable interlocking concrete pavement system was applied to control rainfall runoff, reducing rainfall runoff by 30%–65% during various storm events and minimizing the chances of waterlogging in an urbanized area.[146]

Figure 9: Permeable Road in Tong Wei Tah Street, Kuching City, Sarawak, Malaysia

Source: D.Y.S. Mah et al. 2022. Testing the Concept of Mitigating Urban Flooding with Permeable Road: Case Study of Tong Wei Tah Street, Kuching City, Sarawak, Malaysia. *Trends in Sciences*. 19(15). p. 5592.

In Malaysia, a permeable road consisting of a permeable pavement layer and underground storage was built along Tong Wei Tah Street to mitigate urban flooding issues prevalent in Kuching City, Sarawak. As flooding often occurred because of a backwater effect in the drainage system, the permeable road exhibited the capability to absorb all the out-of-drain floodwaters, leaving no water on the street. Modeling efforts demonstrated that the floodwater hydrographs in the drain rose and fell within 7 hours, while the underground storage was filled and drained within 13 hours.[147]

(iii) Vegetated swales. Vegetated swales—also known as bioswales—are sustainable stormwater management features designed to manage and treat stormwater runoff (Figure 10). These swales are characterized by their broad, shallow channels, often planted with a diverse selection of native, water-resistant plants that have a high potential for removing pollutants. These swales effectively improve water quality in urban areas and help prevent soil erosion and sedimentation, thereby reducing the risk of flooding.[148]

---

[145] USGS. 2019. *Evaluating the potential benefits of permeable pavement on the quantity and quality of stormwater runoff.*

[146] MDPI. 2018. *Rainfall runoff mitigation by retrofitted permeable pavement in an urban area.*

[147] D.Y.S. Mah. et al. 2022. Testing the Concept of Mitigating Urban Flooding with Permeable Road: Case Study of Tong Wei Tah Street, Kuching City, Sarawak, Malaysia. *Trends in Sciences*. 19(15). p. 5592.

[148] PennState Extension. 2022. *Roadside Guide to Clean Water: Vegetated Swales.*

Figure 10: Vegetated Swales Near Rivers

Source: Permaculture research institute. 2015. *Understanding Water Part 2: Working with Flow.*

In Viet Nam's Ho Chi Minh City, officials are exploring ways to optimize the design of Go Vap Cultural Park to create a floodplain park. Its distinctive slope—held together by mangroves and other plants—helps in containing flood levels during storms. Smaller cities like Hue City are experimenting with water-sensitive urban design tools and new types of edge treatments to improve water quality in the Lap River. Vegetated swales are used to treat runoff from nearby streets and foliage to cover the hard embankments bounding the river. Also, Hoi An City is reinforcing its water supply by protecting the Lai Nghi Reservoir from saltwater intrusion. To protect the coastal dunes, planners are defining a coastal zone where vegetation will be planted as a buffer against erosion. ADB estimated that up to $2 million worth of damage because of coastal flooding will be averted each year.[149]

(iv) Algal turf scrubbers. These are water filtering systems designed to remove pollutants from water, particularly nutrients like nitrogen and phosphorus (Figure 11). Algal turf scrubber (ATS) systems use dense mats of simple algae—often referred to as algal turf—to facilitate the biological uptake of these nutrients from polluted water. Such systems also help in recycling nutrients as the harvested algae biomass can be repurposed, such as for biofuel production, organic fertilizers, or as a feed source for livestock. ATS systems can be introduced in wastewater treatment plants to help reduce nutrient loads in treated effluent.

Not only can ATS systems clean water to a high quality, but they can also remove carbon dioxide from the atmosphere by capturing solar energy at rates 10 times that of agriculture and 50 times that of forestry.[150] HydroMentia Incorporation of Ocala, Florida, has designed, constructed, and operated several full-scale systems of up to 40 million gallons per day for aquaculture, agriculture, and municipal nonpoint source treatment. With advanced design and biomass recovery methods, ATS can cost-effectively remove nutrient pollutants from impaired surface waters and provide a valuable tool to meet total maximum daily load (TMDL) requirements.[151]

---

[149] Prevention Web. 2022. *Sink or swim: Southeast Asian cities turn to nature solutions for climate resilience.*
[150] W. Adey and J. Bannon. 2008. *Algal turf scrubbers: cleaning water while capturing solar energy.*
[151] HydroMentia. Algal Turf Scrubber. Results.

Figure 11: Algal Turf Scrubber System, United States

Source: HydroMentia. n.d. *Algal Turf Scrubber.*

(v) Riparian buffers. The use of riparian buffers to maintain water quality in streams and rivers is a forest and conservation management best practice in many countries and is mandatory in some areas. Riparian buffers are vegetated—often forested—areas next to streams, rivers, lakes, and other waterways protecting aquatic environments from the impacts of surrounding land use (Figures 12 and 13). Healthy riparian buffers offer many benefits, including maintaining water quality in waterways by protecting streams from non point sources of pollution such as surrounding agricultural activities; riparian vegetation cover provides a barrier between sediments and pollutants such as nitrates and phosphates, which are washed from the land and water bodies. Riparian buffers are usually planted with native species across several zones to increase their effectiveness and require very little maintenance to perform their function. In addition to their resilience and water quality benefits, riparian buffers also provide important habitats to both aquatic and terrestrial species; temperature moderation from shading creates an important aquatic habitat—especially for fish and insect life—protecting from extreme temperatures (footnote 143). During flood events, riparian vegetation slows run-off by absorbing excess water, reducing peak flow, and helping to mitigate potential flood damage downstream. Some studies have shown that riparian buffers can help to reduce the amount of sediment reaching streams by as much as 80%.[152]

There is multiple evidence globally showcasing how riparian buffers helped in preserving and improving water quality. The Snow Creek Stream Environment Zone in North Lake Tahoe, Placer County, California, was a restoration project involving a brownfield site that had previously been used as a concrete plant since the 1950s. The project restored sensitive environmental areas, provided stormwater treatment, and constructed a multi use trail connecting two existing trails providing recreational benefits. The project revegetated the restored area with native wetland and upland plant species using plants from the undeveloped portion of the project area. It restored 1.25 hectares of riparian area and reestablished about 0.10 hectares of wetlands to mitigate the disturbance caused by the concrete plant (footnote 143).

---

[152] UN Environment-DHI et al. 2018. *Nature-based solutions for water management: a primer.*

Figure 12: Riparian Buffers

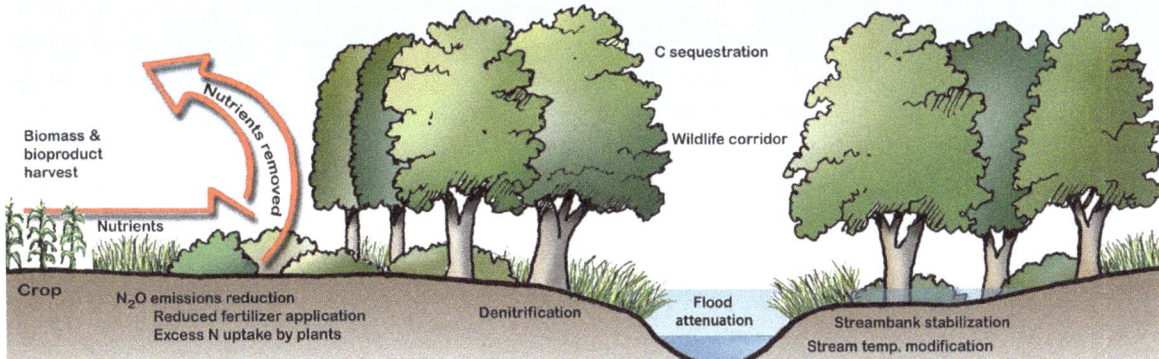

C = carbon, N = nitrogen, N$_2$O = nitrous oxide.

Source: University of Wisconsin-Madison Division of Extension. 2021. *Riparian Buffers—Agroforestry for Any Property.*

Figure 13: Riparian Areas in Snow Creek Stream Environment Zone, California, United States

Source: University of Wisconsin-Madison Division of Extension. 2021. *Riparian Buffers—Agroforestry for Any Property.*

(vi) "Room for the River." In the 1990s, parts of the Netherlands suffered severe flood damage because of the decreasing capacity of flood plains and rivers and rising water levels because of more frequent and heavier rainfall. Because of these extremely high water levels, the Government of the Netherlands started the Room for the River program in 2007, which became the new starting point for the flood protection approach in river areas in the country (Figure 14). The Room for the River program encompassed four rivers: the Rhine, the Meuse, the Waal, and the Ijssel. There were over 30 projects under the program, most of which were completed at the end of 2018.[153]

Room for the River focuses on managing and restoring the natural flow and course of rivers to reduce the risk of flooding and enhance the ecological health of river ecosystems. The projects managed higher water levels in rivers by lowering the levels of flood plains, creating water buffers, relocating levees, increasing the depth of side channels, and constructing flood bypasses. Properly designed Room for the River projects can help prevent erosion along riverbanks, reducing sediment runoff into the river. Pollutants that may be present in the floodwater can be diluted and dispersed, reducing the concentration of pollutants downstream and minimizing their impact on river ecosystems and downstream communities. While Room for the River primarily addresses flood risk management, it can be integrated into a broader river management strategy that includes pollution prevention and control as a key component.[154]

Figure 14: Room for the River in the Netherlands

Source: Dutch water sector. 2019. *Room for the River Programme.*

4.2 Policy Measures

Policy measures to address river pollution typically involve the development and implementation of rules, regulations, laws, or economic incentives aimed at preventing, reducing, or mitigating pollution in rivers. Usually stemming from policymakers in the central government, policy measures to combat river pollution come in a range of complexity and effectiveness.[155]

---

[153] Dutch Water Sector. 2019. *Room for the River Programme.*

[154] MDPI. 2018. Riparian Buffers. *Room for Rivers: Risk Reduction by Enhancing the Flood Conveyance Capacity of The Netherlands' Large Rivers.*

[155] D. Kaczan et al. 2022. *Eco-compensation underpins a greener future for China.* World Bank Blogs.

Two broad categories of policy instruments are used to achieve environmental objectives and control pollution and other externalities. The first is **regulatory standards and controls**, which are often referred to as **command and control policies**. Using these instruments, governments can directly regulate and control the actions of firms or individuals through instruments like bans, standards, penalties, and quotas.[156]

The second type involves **economic incentives,** which are sometimes referred to as **market-based approaches**. Unlike the more rigid command and control approach, economic incentive instruments discourage pollution with monetary incentives (penalties and rewards) such as taxes, fees, or subsidies.[157] By aligning economic interests with environmental goals, these market-based approaches can encourage stakeholders to reduce pollution, adopt more sustainable practices, and protect river ecosystems while also benefiting from cost savings, financial rewards, and market advantages for environmentally responsible actions. In 2021, ADB published a report on Market-Based Approaches for Environmental Management in Asia, including a section on the use of such incentives for water management, and making recommendations on how market-based instruments can be used for more efficient and effective environmental management in Asia.[158] This chapter highlights four economic incentive policy measures for combating river pollution and their applicability to Southeast Asia: (i) payment for ecosystem services (PES); (ii) water quality credit trading; (iii) water pollution taxes; and (iv) extended producer responsibility (EPR). These policy measures have been selected based on their effectiveness at mitigating pollution through mechanisms such as providing economic incentives for ecosystem protection, making polluters pay following the "polluter pays" principle, and encouraging information exchange and capacity building.

4.2.1 Payment for Ecosystem Services

Overview

Ecosystems are essential to civilization and provide society with a wide range of benefits or services including reliable flows of clean water to productive soil and carbon sequestration. People, businesses, and societies rely on these ecosystem services for raw material inputs, production processes, and climate stability. However, many of these ecosystem services are either undervalued or have no financial value, with the result being many ecosystem structures and functions are being undercut and over 60% of the environmental services studied are being degraded faster than they can recover.

PES is a market-based environmental policy instrument that primarily involves offering economic incentives to foster more efficient and sustainable use of ecosystem services. The key characteristic of PES deals is that the focus is on maintaining a flow of a specified ecosystem service—such as clean water, biodiversity habitat, or carbon sequestration capabilities—in exchange for something of economic value. This means that the payment causes the benefit to occur where it would not have otherwise. These schemes provide a new source of income for land management, restoration, conservation, and sustainable-use activities, and by this have significant potential to promote sustainable ecosystem management. PES can therefore support the important aim of halting and reducing the rate of biodiversity loss.[159]

An ecosystem service is any positive benefit that wildlife or ecosystems provide to people. The four main categories of ecosystem services include the following:

(i)   **Provisioning services or environmental goods**: Any type of benefit to people that can be extracted from nature, e.g., supply of food, water, and timber.

(ii) **Regulating services**: The benefit provided by ecosystem processes that moderate natural phenomena, e.g., the regulation of air quality, climate, and flood risk.

---

[156]   Outlier. 2022. *A command-and-control approach to reduce pollution.*
[157]   W. K. Jaeger. 2005. Environmental Economics for Tree Huggers and Other Skeptics.
[158]   ADB. 2021. Greening Markets: Market-Based Approaches for Environmental Management in Asia. Manila
[159]   UNEP. 2008. *Payments for ecosystem services: getting started.*

(iii) **Cultural services**: A nonmaterial benefit that contributes to the development and cultural advancement of people, e.g., opportunities for recreation, tourism, and education.

(iv) **Supporting services**: Essential underlying natural processes that sustain basic life forms, e.g., soil formation and nutrient cycling.[160]

PES schemes occur when the beneficiaries or users of an ecosystem service make payments to the providers of that service. In practice, this may take the form of a series of payments in return for receiving a flow of benefits or ecosystem services. The fundamental principle is that whoever provides a service should be compensated for doing so. A widely quoted definition of PES is as follows (Figures 15 and 16):

(i) It is a voluntary transaction.

(ii) It involves a well-defined ecosystem service (or a land use likely to secure that service).

(iii) The service is "bought by" at least one ecosystem service buyer.

(iv) It is "bought from" at least one ecosystem service provider.

(v) The transaction occurs only if the service provider secures ecosystem service provision (conditionality).[161]

A watershed—simply defined as an area of land that drains into a body of water such as a lake, river, or stream[162]—protects freshwater resources by preventing pollution at its source. The extent of river pollution is closely related to the health of watersheds; because water flows downhill, activities that take place in residences and businesses can influence the water quality of rivers.

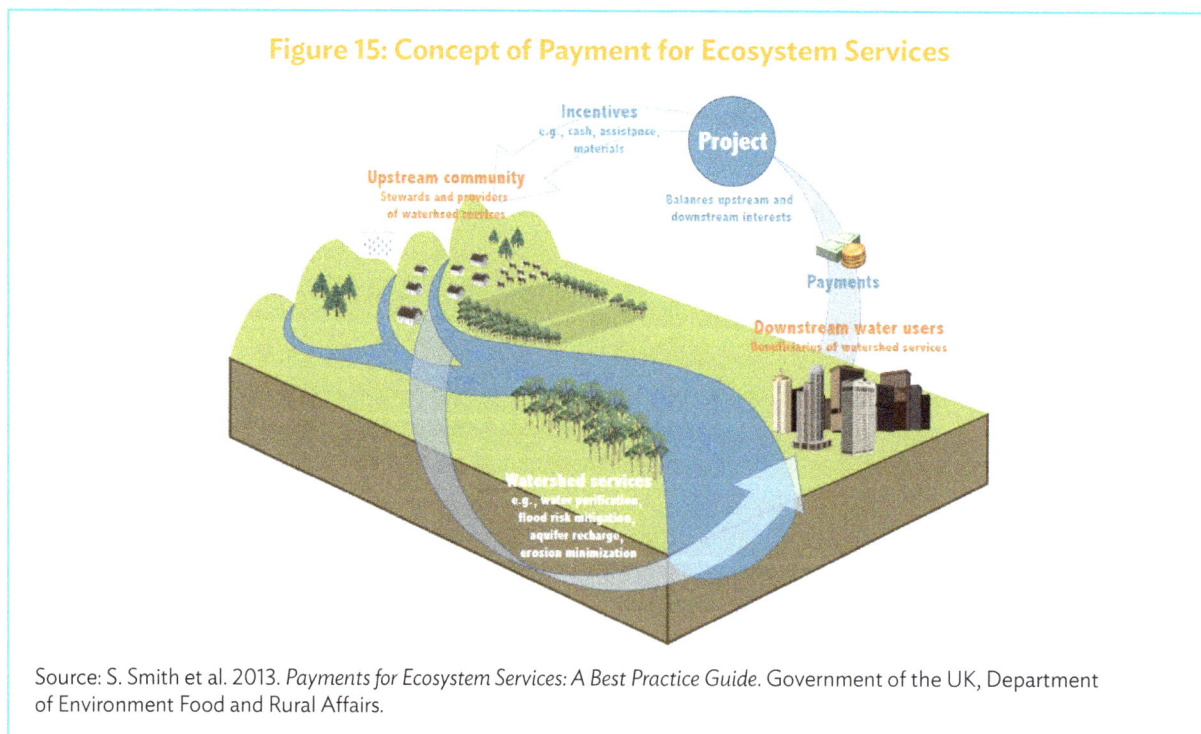

Figure 15: Concept of Payment for Ecosystem Services

Source: S. Smith et al. 2013. *Payments for Ecosystem Services: A Best Practice Guide*. Government of the UK, Department of Environment Food and Rural Affairs.

---

[160] The National Wildlife Federation. *Ecosystem Services*.

[161] E. Fripp. 2014. *Payments for Ecosystem Services (PES): A practical guide to assessing the feasibility of PES projects*. CIFOR.

[162] L. Quillen. 2013. *Protecting the watershed protects drinking water*. Carry Institute of Ecosystem Studies.

Figure 16: Example of Structure for Payments under Payment for Ecosystem Services

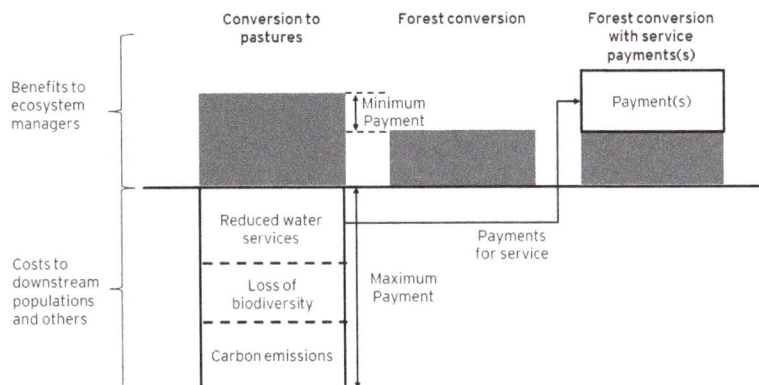

Source: E. Fripp. 2014. *Payments for Ecosystem Services (PES): A practical guide to assessing the feasibility of PES projects.* CIFOR.

Background

The early mentions of ecosystem and environmental services during 1975–1979 refer to climate regulation, flood control, pollination, and biodiversity, but use the concept more as an outreach tool rather than assigning systems monetary value. Throughout the 1990s, the increased conversation around ecosystem services led to its mainstreaming in the literature, with significantly growing interest in methods to estimate economic value.[163] In 2024, PES has been widely recognized as a successful policy tool for natural resource management in over 60 countries. These programs have been implemented for various ecosystem services globally, including biodiversity conservation, watershed services, carbon sequestration, and scenic beauty. The total annual payment amount for PES programs worldwide exceeds $36 billion. Costa Rica, East Africa, Europe, and the PRC are some of the leading countries and regions that have implemented PES.[164] Payments for watershed services exist in India, the US, Bolivia, Costa Rica, Ecuador, Mexico and South Africa (footnote 159).

In Asia, the PRC leads the use of ecological compensation ("eco-compensation"), which refers to fiscal transfers for environmental and natural resources management. While the concept of eco-compensation encompasses PES, it goes beyond PES to include compensation for regulatory takings, direct government-to-government transfers, and frameworks for cooperation.[165] The Yangtze River Protection Law—which came into force in March 2021— represents a landmark legislation to strengthen environmental protection and ecological restoration of a specific river basin in the PRC. The law incorporates specific provisions on the application of ecological compensation policy measures, including (i) establishing a compensation system for ecological protection, (ii) increasing financial transfer payments to compensate areas of ecological importance (e.g., sources of the Yangtze River mainstream and its major tributaries) and key water conservation areas in the upper reaches, and (iii) mandating the development of specific market-based measures to support policy reforms.[166]

---

[163] E. Gómez-Baggethun et al. 2009. *The history of ecosystem services in economic theory and practice: From early notions to markets and payment schemes. Ecological Economics.* Article in Press. 7 November. 03542.

[164] UNDP. 2023. *Promote payment mechanism for natural ecosystem services in Viet Nam.*

[165] World Bank. 2022. *Publication: Ecological Compensation in China: Trends and Opportunities for Incentive-Based Policies towards a Greener China.*

[166] J. Turpie et al. 2008. The Working for Water Programme: Evolution of a Payments for Ecosystem Services Mechanism that Addresses Both Poverty and Ecosystem Service Delivery in South Africa. *Ecological Economics.* 65(4). pp. 788–798.

Implementation

A PES program or scheme is typically developed in one of two ways: either when at least one group of stakeholders recognizes a noticeable depletion in resources leading to genuine demand or when a specific goal is identified—often related to the protection or management of national resources—prompting the introduction of a PES system to establish a market for the service (footnote 166).

The design and implementation of a PES scheme can be divided into five broad phases (Figure 17).

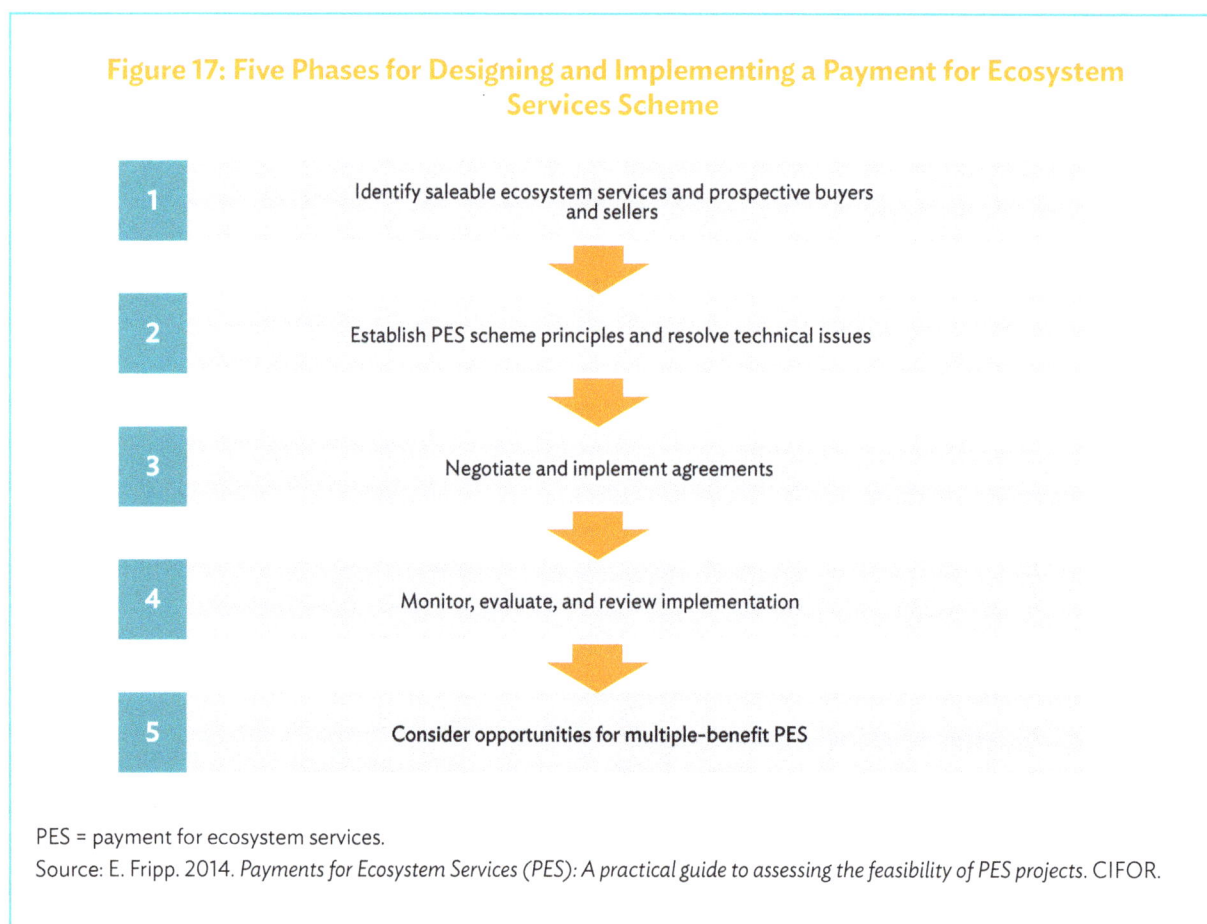

Figure 17: Five Phases for Designing and Implementing a Payment for Ecosystem Services Scheme

1	Identify saleable ecosystem services and prospective buyers and sellers
2	Establish PES scheme principles and resolve technical issues
3	Negotiate and implement agreements
4	Monitor, evaluate, and review implementation
5	Consider opportunities for multiple-benefit PES

PES = payment for ecosystem services.
Source: E. Fripp. 2014. *Payments for Ecosystem Services (PES): A practical guide to assessing the feasibility of PES projects*. CIFOR.

Four principal groups are typically involved in a PES scheme: (i) buyers, (ii) sellers, (iii) intermediaries, and (iv) knowledge providers.[167] The roles of governments in PES schemes are evolving in three distinct ways:

(i) **Buyer**: Government as the buyer of ecosystem services, typically as a strategy to replace or complement government regulation.

(ii) **Regulator**: Government as the regulator, mobilizing private demand for ecosystem services through environmental compliance rules or setting up cap-and-trade systems.

(iii) **Enabler**: Government as an enabler, facilitating the growth of private voluntary transactions.[168]

---

[167] More details on the four principal groups involved in PES are available in Appendix 4.
[168] ADB. 2011. *Buyer, Regulator, and Enabler. The Government's Role in Ecosystem Services Markets*.

Multilateral organizations can play important roles in PES schemes by providing funding for the implementation of such schemes or serving as intermediaries (e.g., for administrative facilitation of transactions)[169] or knowledge providers (e.g., technical assistance, policy advice).[170] There are three key types of payment arrangements for PES schemes:

(i) **Public payment schemes**: These involve the government paying land or resource managers to enhance ecosystem services on behalf of the wider public.

(ii) **Private payment schemes**: Also known as self-organized private deals. In these schemes, beneficiaries of ecosystem services contract directly with service providers.

(iii) **Public–private payment schemes**: These draw on both government and private funds to compensate land or other resource managers for the delivery of ecosystem services.[171]

Perrier Vittel's payments for water quality in France is an example of a self-organized deal. Perrier Vittel (now owned by Nestlé)—a producer of natural bottled mineral water—discovered it would be more cost-effective to invest in conserving the farmland surrounding their aquifers than to build a filtration plant to address water quality issues found in 1990. They then purchased 240 hectares of sensitive habitat and signed long-term conservation contracts with local farmers. Farmers in the Rhine-Meuse watershed in northeastern France received compensation to adopt less intensive pasture-based dairy farming, improve animal waste management, and reforest sensitive filtration zones.

The Public Redistribution Mechanism in Paraná, Brazil, offers an example of a public payment scheme. The state allocated funds to municipalities to protect forested watersheds and rehabilitate degraded areas. In Paraná as well as Minas Gerais, 5% of the revenues received from the circulation of goods and services—an indirect tax charged on the consumption of all goods and services—is distributed either to (i) municipalities with conservation units or protected areas or (ii) municipalities that supply water to neighboring municipalities. The state allocates more revenues to those municipalities with the greatest amount of area under environmental protection (footnote 159).

Applicability to Southeast Asia

Viet Nam has been a leader in piloting the Payments for Forest Environmental Services scheme in Southeast Asia. In 2010, Viet Nam adopted a countrywide PES decree to scale up activities from early models in Lam Dong and Son La—both upland forest areas—around service provision and building on the Viet Nam Forestry Development Strategy 2006-2020 and previous approaches to incentivize forest restoration.[172] The scheme requires water supply, hydropower, and tourist companies to pay fixed rates to environmental services providers—which include state companies and villagers—as they conserve forests for watershed protection and landscape aesthetics. Although Viet Nam has implemented "PES-like" initiatives for marine and wetland ecosystems, there is no comprehensive PES for these environments. Some practical applications include entrance fees for protected marine and wetland areas, co-management models for fisheries resources, and the development of environmentally friendly aquaculture methods in coastal areas with the support of international organizations and related sectors (footnote 164).

---

[169] W. Robichaud. 2014. *Motivation for payments for ecosystem services in Laos: The essential alignment.* Working Paper 142. CIFOR.
[170] ADB. 2010. *Payments for Ecological Services and Eco-Compensation: Practices and Innovations in the PRC.*
[171] S. Smith et al. 2013. *Payments for Ecosystem Services: A Best Practice Guide.* DEFRA.
[172] M. Sommerville. 2016. *Mangrove Payment for Environmental Services in Vietnam: Opportunities & Challenges.* USAID.

In the Lao PDR, there has been motivation to develop PES schemes in the country, with the Nam Theun 2 hydropower project developed in the 2000s being broadly promoted and characterized as a "user pays" PES scheme—or at least a PES-like scheme—by its promoters and developers, the World Bank and Nam Theun 2 Power Company Limited. Some revenues earned from sending reservoir water through the turbines are spent to protect the source of that water, the Nakai-Nam Theun National Protected Area (footnote 169). In Cambodia, there is no legal basis for PES but the idea of environmental services has been featured in key policy documents.[173] In Thailand, there are several activities involving payments for the provision of activities or environmental services, but these are missing many elements that would qualify them as a PES project. Others are mainly at the design stage or the initial stages of implementation.[174]

Challenges and key success factors

In the GMS, the concept of PES has gained attention as a cost-effective and innovative means of promoting sustainable environmental management while improving livelihoods. Yet, a comparison of schemes highlights emerging concerns over equity for participants, the financial sustainability of initiatives, and the need to measure environmental and social outcomes. A series of studies by the Center for International Forestry Research of PES schemes in Cambodia, the Lao PDR, Thailand, and Viet Nam found that far from being market-driven, most PES initiatives are primarily funded by government, donor, and civil society organizations.[175]

Four key factors are essential for the successful implementation of a PES scheme:

(i) **Proper valuation of ecosystem services**: This involves quantifying the impact and conducting economic valuation.

(ii) **Strong legal, policy, and institutional frameworks:** These are crucial, especially in complex environments with intricate ecosystem services and multiple actors with differing agendas.

(iii) **Effective stakeholder organization**: All stakeholders should be well-organized and well-informed about the working mechanisms of PES to ensure that any potential negative impacts or concerns are addressed during early stage planning.

(iv) **Sustainable and sufficient financing**: A successful PES scheme must create a win-win opportunity for both the service provider and the buyers. The buyer should cover the cost of provision, which must (i) be lower than any alternative method the buyer might use to obtain the same service and (ii) be sufficient to make alternatives—such as land conversion—less economically appealing to landowners. This ensures that the incentive to provide the ecosystem service remains intact even in the face of competing land uses.

These factors are especially relevant when operating in a complex environment, such as one with complex ecosystem services within a highly political economy involving many actors with differing agendas. PES often fails because the legal and institutional framework is often poorly defined, and land tenure is frequently unclear or insecure (footnote 161).

[173] S. Milne and C. Chervie. 2014. *A review of payments for environmental services (PES) experiences in Cambodia.* Working Paper 154. CIFOR.

[174] O. Nabangchang. 2014. *A review of the legal and policy framework for payments for ecosystem services (PES) in Thailand.* CIFOR.

[175] A. Hasan. 2014. *Along the Mekong, conservation-payment schemes are a study in contrasts.* CIFOR-ICRAF.

Box 4 presents a successful case study of PES instruments in protecting watersheds and biodiversity in Viet Nam.

Box 4: Protecting Watersheds in Viet Nam through Payment for Ecosystem Services

Background of the project

The degradation of watersheds because of overexploitation has been an alarming phenomenon in Viet Nam for decades. To address this problem, the government established a framework for pilot projects that actively involved local communities in protecting watersheds in Lam Dong and Son La provinces in 2008. It implemented payment for forest ecosystem services schemes, which required beneficiaries of watershed services—such as hydropower plants and water distributors—to compensate suppliers of these services. The country developed a relatively complete legal framework and institutional arrangements that enabled the nationwide implementation of the payment for watershed services program in 2011. From 2008 to 2016, 42 out of 63 provinces established Provincial Forest Protection and Development Funds for more than 100,000 forest service providers who protect 3.3 million hectares (27%) of the total forested area.

Impacts of the project

The program is a successful case study of using market-based instruments in protecting watersheds and biodiversity in Southeast Asia. It helped to reduce the government's financial burden for forest conservation and protection, improved economic conditions, and created sustainable livelihoods for communities protecting and managing the forests.

As of mid-2016, $257 million was collected for the protection of 5.6 million hectares of forest, while more than 500,000 households—made up of mainly ethnic minorities and poor households living in the forested area—received money from the scheme. The payment for forest ecosystem services policy has contributed to the increase in household income by about D3.9 million (about $170) per household per year in the watershed area of Dong Nai River. This income has been used for basic household consumption or investments in agricultural activities. From 2008 to 2015, the forest area and forest cover rate increased in general across the country—particularly in Son La province—but slightly decreased in Lam Dong. As of 2015, 41,649 households in Son La and 17,073 households in Lam Dong were provided with and paid for forest ecosystem services.

Overall, the payment for forest ecosystem services program created incentives for forest resource users to conserve and manage natural resources more sustainably. At the same time, it reduced the burden per household by distributing costs for forest environmental services among stakeholders. The payment for forest ecosystem services policy ensured that service providers, especially vulnerable people—poor people, older people, or landless households—benefited from it. However, the method of calculating payments under the program could create unequal outcomes among service providers. Communities with watersheds with a higher percentage of forest area receive a smaller average payment rate per hectare, while those with a lower percentage of forest area receive a larger payment under the program.

Challenges and key success factors

The success of the payment for forest ecosystem services policy for water management in Viet Nam can be attributed to three main factors:

(i) **Enabling policy framework.** The pilot programs in Lam Dong and Son La Provinces led to the establishment of the payment for forest ecosystem services legal framework and the development of a governing system at different levels in the environmental management structure in Viet Nam. This is

continued on next page

Box 4 *continued*

an important milestone in the policy agenda of adopting market-based approaches for environmental management in the country.

(ii) **Public acceptance.** After 10 years of implementation, the policy has gained support from local communities and pay enterprises because of manageable costs and benefits to service providers, especially vulnerable groups.

(iii) **Technical and financial support**. International organizations provided technical and financial assistance, shared experiences, and support in formulating the practical and theoretical framework during all stages.

A key feature of Viet Nam's payment for the forest ecosystem services program is that forest land use rights were devolved from the state directly to households. This type of reform is distinct from other countries that devolved rights from state to community in the form of community-based forest management. While individual household contracts for service providers imply high transaction costs, the program demonstrated the high administrative capacity of intermediary agencies to ensure well-defined terms for each contract, an efficient payment scheme, and effective monitoring and evaluation. With community contracts to lower transaction costs, coupled with government capacity, such payment schemes could be even more attainable for other countries and contexts. The lessons learned from forest environmental services in Viet Nam provide a basis for scaling up similar mechanisms for other ecosystems, including watersheds.

Source: P. Nam et al. 2021. Protecting Watersheds in Viet Nam through Payment for Ecosystem Services. Development Asia.

4.2.2 Water Quality Credit Trading

Overview

Water quality credit trading is a market-based approach to achieving water quality improvements. It allows a pollution source to control pollutant levels beyond required limits and sell credits to another source, enabling credit trades to meet regulatory requirements. Regulatory standards can be incrementally raised, making this method effective in managing pollutants from multiple sources that collectively affect water quality.

Water quality credit trading offers flexibility in the timing and extent of technology adoption by facilities while attracting private investment for innovative, cost-effective water quality solutions. Additionally, it promotes voluntary participation of nonpoint sources within the watershed. For instance, nonpoint source projects like streamside buffers or conservation tillage, integrated into water quality trades, can yield profits while providing environmental benefits such as carbon emissions reduction, flood risk mitigation, streambank stabilization, and wildlife habitat preservation.[176]

Background

In 2008, the World Resources Institute assessed water quality credit trading programs worldwide. They identified 57 programs. Of these, 26 are active, 21 are under consideration or development, and 10 are inactive. Of the programs identified, all but six are in the US. The six trading programs that are not in the US comprise the following:

---

[176] EPA. *Water Quality Trading.*

(i)   New Zealand, the Lake Taupo Nitrogen Trading Program

(ii)   Australia, Hunter River Salinity Trading Scheme

(iii) Australia, South Creek Bubble Licensing Scheme

(iv)   Australia, Murray Darling Basin Salinity Credits Scheme

(v)   Australia, the Moreton Bay Nutrient Trading Scheme

(vi)   Canada, South Nation River Watershed Trading Program[177]

A study in 2023 confirmed that the four countries, Australia, Canada, New Zealand, and the US, remained the only countries globally with water quality credit trading programs.[178]

Implementation

There are six key considerations associated with the implementation of a water quality credit trading scheme (footnote 177):

(i)   **Policy drivers.** The primary policy driver for all water quality trading programs has been the implementation or forthcoming implementation of nutrient caps that limit pollutant discharges.

    (a)   **US**—The Clean Water Act provides the foundation for point-source nutrient caps by requiring states to adopt water quality standards for various pollutants. Violation of these standards may result in a TMDL being developed for the waterbody. The pollutant limit allocated to point sources under a TMDL—or "wasteload allocation"— forms the basis of a water quality-based effluent limit that is placed in a regulated facility's permit. These permit limits—or threat of permit limits—have driven the development of a large number of water quality trading programs in the US.

    (b)   **New Zealand**—Given that the Resource Management Act grants regional governments the authority to make resource management decisions, the Waikato Regional Council has imposed nitrogen discharge caps on all sources in the Lake Taupo catchment.

    (c)   **Canada**—The provincial Ministry of Environment guidelines are the driver for the South Nation River Watershed Trading Program in Ontario, Canada. The ministry is responsible for water quality and sewage treatment plant licensing in Ontario. It stipulates that no new pollutant discharge is allowed in a watershed if water quality guidelines are exceeded.

    (d)   **Australia**—The Hunter River Salinity Trading Scheme in New South Wales is driven by salinity concerns for the Hunter River. The New South Wales Environmental Protection Agency set a numeric salinity goal for the river, with the major point sources holding an environmental protection license to discharge. Similarly, the agency created a total pollutant load limit for nutrients in South Creek and allowed the affected sewage treatment plants to trade to stay within that limit.

(ii)   **Water quality cap allocation**. Once a watershed water quality cap has been established, the cap must be allocated among all regulated entities. Pollutant caps for point sources are generally allocated based on regulatory numeric effluent concentration limits for a given pollutant. To facilitate trading, effluent pollutant concentration limits are often translated into an annual discharge limit expressed as a unit of mass over time (e.g., kilos per year). The annual discharge limit is based on the numeric effluent concentration limit and an annual facility flow volume.

---

[177]   M. Selman et al. 2009. *Water Quality Trading Programs: An International Overview*. WRI Issue Brief, Number 1.
[178]   Zapata et al. 2023. Water Quality and Pollution Trading: A Sustainable Solution for Future Food Production.

(iii) **Establishment of nonpoint source baselines.** As nonpoint sources are typically not regulated, their baseline nutrient discharges must be established before they can generate and trade any nutrient reduction credits. Establishing baselines ensures that credits generated by nonpoint sources are "additional" water quality improvements that would not otherwise have taken place.

(iv) **Nonpoint source nutrient loss and reductions calculations.** Because reductions in nutrient losses from nonpoint sources are difficult to measure, program designers must identify the measurement or estimation approach they will use to determine the nutrient losses and reductions from these sources. Three common approaches are as follows:

   (a) **Direct measurement through monitoring.** This approach uses direct measurements based on in-field samples to determine the nutrient reductions that result from the implementation of a control measure.

   (b) **Site-specific calculations.** This approach uses established calculation methodologies to estimate nutrient losses and reductions from nonpoint sources, considering site-specific variables such as soil type, slope, and fertilizer application rate.

   (c) **Predetermined nutrient reductions for practices regardless of location or other site-specific characteristics.** This approach assigns a predetermined reduction credit for each practice based on an estimated average nutrient reduction. These credit values are generally derived from scientific literature or watershed-level modeling and do not change across the watershed or region.

(v) **Trading ratios.** Trading ratios are frequently used to account for factors such as uncertainty in reduction estimates, creating equivalency among multiple pollutants, ensuring overall water quality benefits, accounting for the effects of nutrient transport, and mitigating buyer risks. Trading ratios are applied to the estimated nutrient reductions to determine the saleable reduction credit. For instance, a 2:1 trading ratio means that an entity needs to purchase 0.9 kilos of pollutant reductions to offset every 450 grams they discharge above their regulatory limit.

(vi) **Market structure.** Market structure defines how trading will occur and the infrastructure used to support the water quality trading program. Common types of trading are the following:

   (a) **Bilateral trades.** Bilateral trades are characterized by one-on-one negotiations where a price is typically arrived at through a process of bargaining and not simply by observing a market price. The Virginia Water Quality Trading Program has a hybrid bilateral/clearinghouse market structure.

   (b) **Sole-source offsets.** Sole-source offsets occur when sources are allowed to increase nutrient discharge at one point if they reduce their nutrient discharge elsewhere (either on-site or off-site). The Cherry Creek Reservoir Watershed Phosphorus Trading Program and the Chatfield Reservoir Trading Program in Colorado have this market structure.

   (c) **Clearinghouse.** A clearinghouse market is one where a single intermediary links buyers and sellers of credits. The clearinghouse converts a commodity that may have a variable price—such as a nutrient credit—into a uniform commodity.

   (d) **Exchange market.** An exchange market is where buyers and sellers meet in a public forum (e.g., online) with all commodities being equivalent and all prices transparent.

Applicability to Southeast Asia

While water quality credit trading has yet to be applied in Southeast Asia, it has the potential to bring the following benefits:

(i) **Reducing agricultural pollution**. Water quality credit trading could be used to encourage farmers to adopt best practices that reduce agricultural runoff, such as planting cover crops and using buffer strips.

(ii) **Improving wastewater treatment**. Water quality credit trading could be used to encourage businesses and industries to improve their wastewater treatment practices.

(iii) **Protecting water bodies**. Water quality credit trading could be used to protect water bodies such as rivers, lakes, and estuaries by reducing pollution from all sources.

Challenges and Key Success Factors

Five key success factors are required for a water quality trading program:

(i) **Strong regulatory and/or nonregulatory drivers**, which help to create a demand for water quality credits.

(ii) **Minimal potential liability risks** to the regulated community from meeting regulations through trades.

(iii) **Robust, consistent, and standardized estimation methodologies** for nonpoint source actions.

(iv) **Standardized tools, transparent processes, and online registries** to minimize transaction costs.

(v) **Buy-in from local and state stakeholders** (footnote 177)**.**

Box 5 presents a project example of water quality credits trading in the US.

Box 5: Water Quality Credits Trading in Chesapeake Bay, United States

Background of the project

Considered a "national treasure and resource of worldwide significance," the Chesapeake Bay watershed is the largest estuary in the US and one of the largest and most productive in the world. Yet, it has suffered from excess nutrients and sediment for decades.

The Clean Water Act serves as the primary legislation guiding water quality efforts, including those in the Chesapeake Bay watershed. After voluntary attempts at improving water quality failed to deliver results, the act was updated to require states to formulate total maximum daily load requirements, which establishes the maximum allowable pollutant discharge into a water body before it jeopardizes water quality standards. To help organizations cost-effectively meet total maximum daily load allocations, three states (Maryland, Pennsylvania, and Virginia) developed water quality trading programs in the watershed during 2005–2008. This allowed sources with high pollution control costs to purchase credits—or pollution discharge reductions—from sources with lower pollution control costs.

Impacts of the project

The water quality trading scheme in Chesapeake Bay led to several positive outcomes:

(i) **Reduced cost and increased speed of compliance**.

(ii) **Flexibility in growth management.** Trading provided flexibility in meeting regulatory requirements and helped to manage growth and water quality in a capped watershed.

continued on next page

Box 5 *continued*

(iii) **New revenue sources.** The scheme created new revenue sources—especially for farmers—incentivizing them and other nonpoint sources to actively participate in water quality management.

(iv) **Developing green infrastructure valuation.** Water quality trading contributes to improvements in the valuation of green infrastructure, which offers additional benefits like habitat creation, greenhouse gas mitigation, and recreation, beyond just improving water quality.

(v) **Community building**. Water quality trading fostered communication between rural and urban communities, enhancing relationships and collaboration.

Key lessons learned

Successful water quality trading programs rely on strong regulatory drivers to create demand for trading. These programs must be well designed to protect water quality and establish efficient and credible markets. This is especially important as credit buyers may have concerns about the legitimacy of credits, particularly when they come from diffuse sources like agriculture.

Furthermore, stakeholder engagement and buy-in are essential from the program's inception. To ensure cost-effectiveness and encourage participation, transaction costs must be minimized, including expenses related to stakeholder engagement, credit quantification, buyer–seller connections, and verification of credit-generating activities. Developers of such programs should carefully consider these costs in relation to expected credit prices and water quality improvements.

Source: S. Walker. *Water Quality and Agriculture: Water Quality Trading in the Chesapeake Bay Watershed, USA*. Organisation for Economic Co-operation and Development (OECD).

4.2.3 Water Pollution Taxes

Overview

Taxes are a flexible market-based policy instrument that can minimize administrative control costs for achieving a given pollution target. Depending on the specific design features, water pollution taxes support the "polluter pays principle," under which the costs of pollution prevention and control should be reflected in the price and output of goods and services that cause pollution as a result of their production and/or consumption. Water pollution taxes introduce a price signal that helps ensure that polluters take into account the costs of pollution on the environment when they make production and consumption decisions.[179]

There are three main types of water pollution taxes:

(i)   **Direct pollution taxes**, based on pollutant discharge volumes.

(ii) **Indirect pollution taxes**, often in the form of product taxes.

(iii) **Preferential tax policies**, intended to encourage environmentally friendly behavior such as reduced or discounted income tax, value-added tax, and/or consumption tax.[180]

---

[179]   OECD. Taxation and Environmental Policies.
[180]   ADB. 2011. *Market-Based Instruments for Water Pollution Control in the People's Republic of China*.

Benefits of a tax approach include the promotion of innovation (as opposed to static quantity-based regulations) and the generation of revenue, especially since many water quality projects involve substantial investments in infrastructure. Taxes collected are often earmarked for environmental protection and the development of the environmental industry, and these taxes can provide a readily available source of funds for the development of treatment plants, sewer systems, and flow control infrastructure like dams that require significant public financing.[181] Many countries have also initiated tax reductions and rebates to encourage enterprises to conserve resources, control pollution, and support research and development in these areas (footnote 180).

Background

In the 1960s and 1970s, growing environmental concerns regarding water pollution led to the implementation of new environmental regulations in several countries. France has had the Water Act in place since 1964, Germany introduced amendments to the Water Household Act in 1976, and the Netherlands enacted the Surface Waters' Pollution Act in 1969. These measures signaled a shift toward a more proactive approach to water pollution control.

Furthermore, in the 1960s, France began imposing water charges on industrial polluters, and in 1976, Germany passed an Effluent Charge Law. Early advocates of tax-based pollution control systems saw promise in the French and German experiments as potential models for other nations. However, subsequent analysis revealed limitations in these systems. They often operated alongside treatment-based regulations, primarily serving as revenue sources and setting taxes at relatively low levels, which limited their effectiveness in driving significant pollution reduction.

An exception was the Netherlands, where their program levied charges on both direct and indirect sources of pollution and used the revenues to support improved treatment. The program stood out because of its higher fee levels and consistent increases over time, motivating pollution reductions independently. Econometric analysis indicated that the effluent fees—rather than treatment standards—significantly contributed to enhancing water quality.[182]

Over the decades, many countries have implemented pollution taxes aimed at internalizing the environmental costs of pollution and encouraging businesses and industries to reduce their emissions. In 2023, water pollution taxes continue to play a vital role in addressing water quality issues and are particularly common in Organisation for Economic Co-operation and Development (OECD) economies (footnote 179).

Implementation

The implementation of water pollution taxes should adhere to nine general principles:

(i)   Tax bases should be focused on the pollutant or polluting behavior, with few exceptions.

(ii)   The scope of a water pollution tax should ideally match the extent of watershed damage.

(iii)   Tax rates should be proportionate to the watershed harm.

(iv)   The tax must be credible, and its rate should be predictable to encourage environmental improvements.

(v)   Tax revenues can support fiscal consolidation or reduce other taxes.

(vi)   Distributional concerns should generally be addressed through other policy instruments.

(vii)   Competitiveness concerns require careful assessment, and coordination and transitional relief can be effective responses.

---

181   J. Boyd. 2003. *Water Pollution Taxes: A Good Idea Doomed to Failure?* Discussion Paper 03-20. Resources for the Future.
182   M. Andersen. 2001. *Economic Instruments and Clean Water: Why Institutions and Policy Design Matter.* OECD.

(viii) Clear communication is crucial for public acceptance of taxation.

(ix) Taxes may need to be combined with other policy instruments to address specific issues.[183]

Water pollution taxes are usually initiated and advocated at the federal level before being delegated to the state, municipal, or local level, or the responsible person or agency in charge of a specific watershed for policy implementation.

(i) **People's Republic of China:** In January 2018, the PRC introduced the Environmental Protection Tax, covering water pollution, air pollution, noise pollution, and solid waste. A noteworthy shift from the past, all tax revenue now stays at the local level, whereas previously, the central government retained 10% of the fees. This change grants local authorities more control and enforcement capacity regarding the tax. Moreover, local governments have some discretion, allowing them to set tax rates for various pollutants within a range determined by the central government. This flexibility enables them to consider local socioeconomic conditions and the unique situations of polluting enterprises.

(ii) **France:** There are six French water agencies established under the French Water Law of 1964. Their primary mission is to implement water policies that align with national guidelines but are tailored to the basin level. One of the ways they achieve this is by collecting fees (in the form of water pollution taxes) from various sectors of the economy (including industry, households, and agriculture) for the release of pollutants into water bodies.[184]

Applicability to Southeast Asia

While no jurisdiction in Southeast Asia has water pollution taxes implemented, some preliminary work has been done in the realm of environmental taxes in the region. During 2013–2023, Thailand attempted to implement a comprehensive environmental tax reform which has so far been unsuccessful, while Viet Nam has seen more success in this regard. The feasibility and usefulness of implementing a comprehensive environmental tax reform depend on the specific country context. Policymakers need to carefully assess whether a broader process of environmental fiscal reform is both institutionally and politically feasible. The complexity of proposed legislation can involve various stakeholders, both within the government (ministries and other decision-making bodies) and external entities (businesses, civil society) in the decision-making process.[185]

Challenges and key success factors

There are four main challenges in the implementation of water pollution taxes:

(i) **Equity and distributional effects.** The effective incidence of environmental taxes is likely to differ from their formal incidence. What this means is that imposing taxes on polluters could end up having a significant impact on low-income households if the burden is passed on (footnote 183).

(ii) **International competitiveness.** Water pollution taxes could lead to reduced international competitiveness in certain industries of the economy and may even lead to businesses relocating to lower-taxed jurisdictions (footnote 179).

---

[183] OECD. 2011. *Environmental Taxation: A Guide for Policy Makers.*

[184] A. Courtecuisse. 2008. *The role played by the French water agencies.* Agence de l'Eau Artois-Picardie.

[185] J. Cottrell et al. 2017. *Environmental Tax Reform in Asia and the Pacific.* UNESCAP.

(iii) **Political resistance**. The implementation of water pollution taxes faces notable political challenges, as it often clashes with industry players grappling with rising expenses and voters who will experience heightened prices.[186]

(iv) **Monitoring and enforcement**. Water pollution taxes necessitate a strong institutional setup to effectively monitor and enforce compliance with tax regulations.[187]

Policymakers should consider four key success factors:

(i) **Well-designed tax.** Water pollution taxes need to be crafted with a precise focus on the pollutants and polluting behaviors that need to be addressed. This ensures that the tax system accurately targets the sources of pollution, allowing for more effective environmental outcomes.

(ii) **Transparent and predictable framework**. Creating a transparent and predictable taxation framework involves clearly defining how the tax is calculated, when it is due, and how the revenue will be allocated or used. This fosters an environment of trust and stability, encouraging compliance and long-term planning for pollution reduction among businesses and industries.

(iii) **Stakeholder engagement.** Engaging various stakeholders—including businesses, environmental organizations, and communities—in the development and implementation of water pollution taxes ensures that diverse perspectives are considered, builds support for the tax, and addresses potential concerns or objections.

(iv) **Effective monitoring.** Establishing a robust monitoring system to track pollution sources, collect relevant data, and enforce tax payments and penalties ensures that the tax system remains credible and encourages compliance, leading to improved pollution reduction outcomes (footnote 184).

Box 6 presents the impacts of water pollution taxes in the People's Republic of China.

Box 6: Water Pollution Taxes in the People's Republic of China's Environmental Protection Tax Law

Background of the policy

In 2018, the national Environmental Protection Tax (EPT) Law was introduced in the People's Republic of China (PRC), replacing the longstanding pollution discharge fee that had been in place since the 1980s. This marked a significant shift toward a more formal tax regime in environmental management. The underlying rationale for these increases in EPT rates by local governments was to heighten the cost of polluting for companies, thereby creating a stronger incentive for pollution control and reduction.

The EPT Law provides a framework within which local governments can determine their specific tax rates, falling within the upper and lower bounds defined by the law. For example, the tax rates for water pollutants can range from CNY1.4 per equivalent unit to CNY14 per equivalent unit, and individual provinces had the discretion to set their own EPT rates, with some provinces opting to keep their tax rates unchanged. This divergence in tax rate policies reflects the increasingly decentralized nature of environmental governance in the PRC.

continued on next page

---

186 OECD. 2017. *Diffuse Pollution, Degraded Waters: Emerging Policy Solutions.*
187 D. Whittington et al. 2021. *Economic Tools to Help Manage Asia's High Demand for Water Resources.* Development Asia.

Box 6 *continued*

Impacts of the policy

Higher taxes have led to substantial reductions in both chemical oxygen demand and ammonia nitrogen emissions by companies. It is estimated that a one unit tax rate increase (or CNY1 increase per pollutant equivalent) would lead to a 9% reduction in firm-level chemical oxygen demand emissions and a 4% decrease in ammonia nitrogen emissions.

Firms in industries with higher emission intensity tend to reduce emissions and pollutant concentrations more than firms in industries with lower emission intensity when subjected to higher environmental tax rates, indicating a negative relationship between environmental tax rates and firm emissions. Policy effects were more pronounced in developed provinces, with more significant pollutant reductions relative to less developed provinces.

Weaknesses of the policy

(i)   **Reluctance to tax**. Regions with large manufacturing bases may set their tax rates lower to retain the fiscal revenue.

(ii) **Inhibits growth**. The imposition of the EPT significantly reduces the performance of heavily polluting companies in the short run, a mechanism test shows that the EPT has an innovation effect, "forcing" companies to increase their research and development investment and realize transformation and upgrading, which inhibits the growth of corporate performance in the short run.

(iii) **Differing effects across regions.** The regulatory effect of the EPT is more evident in eastern PRC where the institutional environment is more favorable.

CNY = PRC yuan.
Source: Y. Zhang et al. 2023. Can raising environmental tax reduce industrial water pollution? Firm-level evidence from China.

4.2.4 Extended Producer Responsibility

Overview

The OECD defines extended producer responsibility (EPR) as an environmental policy approach in which a producer's responsibility for a product is extended to the post-consumer stage of a product's life cycle. An EPR policy is characterized by two features:

(i)   The shifting of responsibility (physically and/or economically; fully or partially) upstream toward the producer and away from municipalities.

(ii) The provision of incentives to producers to take into account environmental considerations when designing their products.

While other policy instruments tend to target a single point in the chain, EPR aims to incorporate signals about a product's environmental aspects and how it is made across the entire product chain.[188]

---

[188]   OECD. Extended Producer Responsibility.

Background

During 1990–1994, several European countries initiated strategies to improve end-of-life management of products. Germany introduced the first example of EPR in Europe in 1991 with a requirement that manufacturers assume responsibility for recycling or disposing of packaging material they sold. In response, the country set up a "dual system" for waste collection, picking up household packaging alongside municipal waste collections. Today, almost all members of the OECD have established EPR policies as an approach to pollution prevention and waste minimization (Figure 18).[189]

Figure 18: Cumulative Global Adoption of Extended Producer Responsibility Policy, 1970–2015

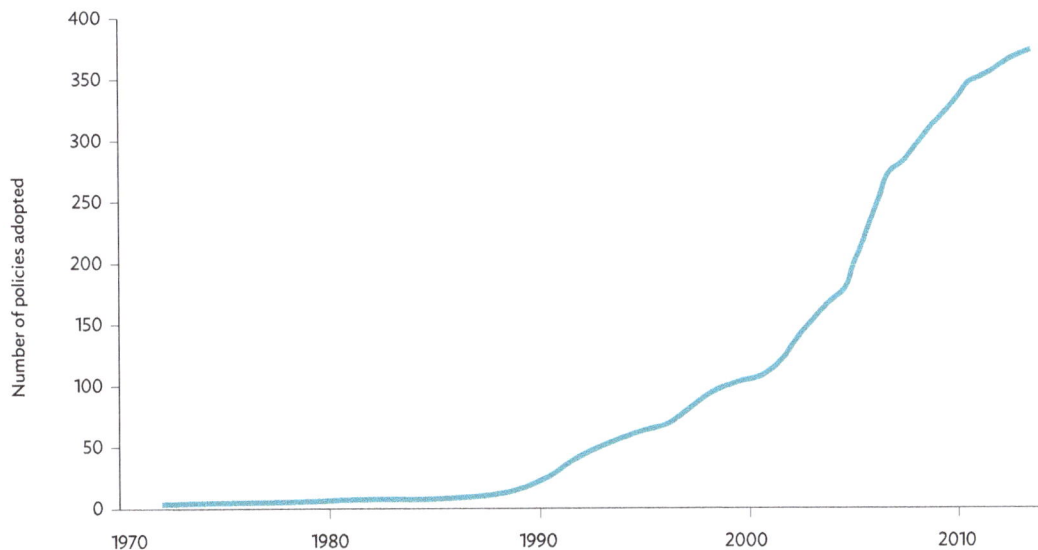

Source: OECD. 2016. *OECD Policy Highlights—Extended Producer Responsibility.*

A survey conducted by the OECD reveals that there are 400 EPR systems in operation as of 2023, of which three-quarters were established since 2001. Legislation has been a major driver, and most EPRs appear to be mandatory rather than voluntary. Small consumer electronic equipment accounts for more than one-third of EPR systems, followed by packaging and tires (each 17%), end-of-life vehicles, lead-acid batteries, and a range of other products. Various forms of take-back requirements are the most commonly used instrument, accounting for nearly three-quarters of those surveyed. While in some cases individual firms have established their systems, in most cases, producers have established collective EPR systems managed by producer responsibility organizations (PROs) (Figure 19).[190]

---

[189] Multi-Material Stewardship Western. *History of EPR.*
[190] OECD. 2016. *OECD Policy Highlights—Extended Producer Responsibility.*

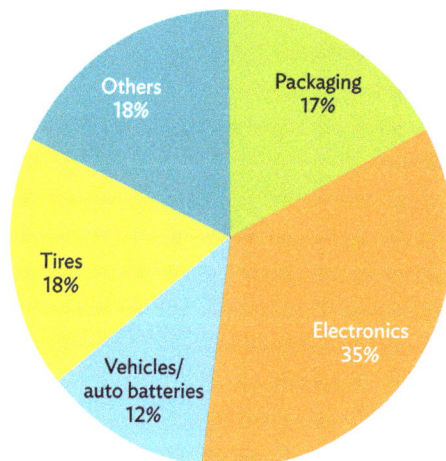

Figure 19: Extended Producer Responsibility by Product Type Worldwide

Packaging 17%
Others 18%
Tires 18%
Vehicles/auto batteries 12%
Electronics 35%

Source: OECD. 2016. *OECD Policy Highlights—Extended Producer Responsibility*.

The landscape of EPR in Asia varies significantly across countries and between OECD and non-OECD members. Industrialized OECD economies like Japan and the Republic of Korea have already established well-established EPR schemes and regulations in place on the key waste streams, supported by a solid monitoring and enforcement framework. The PRC began the promotion of producer responsibility schemes in 2017 and aims to have supporting regulations and rules ready by 2025.[191]

Implementation

Governments at both national and regional levels play a vital role in promoting and facilitating EPR. They achieve this by creating the required policies and legislation and fostering an environment conducive to effective EPR schemes. Additionally, governments can engage and provide support to various stakeholders, including businesses, waste management companies, the informal sector, and the public, to ensure the successful implementation of EPR initiatives.[192]

Policymakers need to address five key considerations to implement a successful EPR scheme for water pollution within the broader legislative framework of water management:

(i) Define EPR within the context of water pollution, considering factors such as initial pollutant concentration, toxicity, and potential transformations after prolonged exposure to water.

(ii) Develop a specific list of substances that should be covered by the EPR scheme. This list should focus on pollutants that persist in water and have adverse effects over time.

(iii) Establish criteria for identifying parties responsible for water pollution, including producers, consumers, and recyclers.

191 *Xinhuanet*. 2017. China unveils extended producer responsibility plan.
192 E. Watkins et al. 2020. *How to Implement Extended Producer Responsibility (EPR)*. WWF.

(iv) Create a methodology to ensure equitable cost-sharing among all stakeholders within the supply chain—including importers—to address pollution-related expenses.

(v)   Establish a PRO responsible for defining the mandates and roles of local, national, and regional EPR organizations. This PRO should oversee and harmonize the implementation of the EPR methodology, recognizing the importance of collaboration among regional organizations because of the transboundary nature of water pollution.[193]

There are four broad categories of EPR instruments at the disposal of policymakers. These typically address specific aspects of waste management, but can be implemented concurrently:

(i)   **Product take-back requirements**. Take-back policies require the producer or retailer to collect the product at the post consumer stage. This objective can be achieved through recycling and collection targets of the product or materials and through incentives for consumers to bring the used product back to the selling point.

(ii) **Economic and market-based instruments.** These include measures such as deposit-refund schemes, advanced disposal fees, material taxes, and upstream combination tax and/or subsidy that incentivize the producer to comply with EPR. In the Republic of Korea, for example, advance disposal fees are imposed on importers and producers of products that are hazardous and more difficult to recycle.

(iii) **Regulations and performance standards such as minimum recycled content**. Standards can be mandatory or applied by industries themselves through voluntary programs.

(iv) **Accompanying information-based instruments**. These policies aim to indirectly support EPR programs by raising public awareness. Measures can include imposing information requirements on producers such as reporting requirements, labeling of products and components, communicating to consumers about producer responsibility and waste separation, and informing recyclers about the materials used in products.

Applicability to Southeast Asia

The Government of Indonesia is aligning itself with other Asian nations by introducing an EPR system as part of its national policy to combat plastic pollution, particularly related to products and packaging. The Ministry of Environment and Forestry Regulation No. 75/2019 (referred to as the "EPR Regulation") and the Roadmap to Waste Reduction by Producers outline the obligation for producers to reduce waste from their products and packaging by 30% by 2029.[194]

Viet Nam introduced a new Law on Environmental Protection and Decree No. 08/2022/ND-CP during 2020-2022, establishing the legal basis for EPR. Decree No. 08/2022/ND-CP outlines the obligations for producers such as registering with authorities, developing and implementing EPR plans, and contributing to waste management and recycling efforts. The decree designates specific product categories for which EPR requirements apply, including electronic products, batteries, packaging materials, and other items with potential environmental impacts.

In Thailand, a collaboration between the Government of Thailand, the private sector, and NGOs in 2018 led to the formation of the Thailand Public-Private Partnership for Plastic and Waste Management, which focused on constructing drop-off points and material recovery facilities in Bangkok City. While EPR is voluntary, the Pollution Control Department under the Ministry of Natural Resources and Environment has legitimized Thailand's Roadmap on Plastic Waste Management Phase 2. This roadmap includes plans to enact an EPR law on packaging by 2027, covering all types of packaging to promote resource circularity.

---

[193] CEFIC. 2023. *Cefic views on Extended Producer Responsibility in water policies.*
[194] WWF. 2022. *Extended Producer Responsibility Guideline on Plastic Products and Packaging for Industries in Indonesia.*

Key challenges of Extended Producer Responsibility implementation

There are four main challenges when implementing EPR:

(i) **Design and governance of extended producer responsibility**. The design of an EPR system involves determining which products or waste streams will be covered, setting targets for recycling and waste reduction, establishing collection and recycling mechanisms, and defining the roles and responsibilities of various stakeholders. Governance is equally important as it involves creating the legal and regulatory framework for EPR, ensuring transparency, and enforcing compliance. Striking the right balance between government involvement and industry cooperation is a key consideration.

(ii) **Competition issues**. EPR can impact market competition, especially when it comes to the financing of recycling and waste management. Producers may pass the costs of EPR programs onto consumers, potentially affecting product prices. Addressing competition concerns involves carefully designing fee structures and ensuring that EPR does not create barriers to entry for new businesses in the recycling and waste management sector.

(iii) **Design for environment incentives**. EPR programs can provide incentives for eco-friendly product design, but setting up effective incentives requires careful consideration. Producers should be motivated to use materials that are easier to recycle, reduce packaging, and create products with longer lifespans. This involves striking a balance between regulatory requirements and market-based incentives.

(iv) **Role of the informal sector**. In many countries, the informal sector, including waste pickers and recyclers, plays a significant role in waste collection and recycling. EPR systems need to consider the inclusion of the informal sector and provide opportunities for formalization and recognition of their contributions. This can involve integrating informal workers into the formal waste management system, ensuring fair compensation, and addressing health and safety concerns (footnote 190).

Box 7 presents an example of EPR application in Japan.

Box 7: The Packaging Recycling Act—Application of Extended Producer Responsibility to Packaging Policies in Japan

Background of the policy

In response to increasing municipal solid waste (MSW) and diminishing space in MSW landfill sites, the Act on the Promotion of Sorted Collection and Recycling of Containers and Packaging (Packaging Recycling Act) was introduced in Japan in December 1995. The legislation was designed to create a new system to promote sorted collection and storing of containers and packaging by municipalities as well as promoting their recycling by business operators. The targets of the act are the containers and packaging that account for a large percentage of MSW, and where recycling technologies are available.

The Packaging Recycling Act requires designated producers to recycle waste packaging that meets specific sorting criteria after being collected in a sorted state by municipalities. To fulfill the recycling obligation, the designated producers can either collect and recycle waste packaging that conforms to the specified sorting standards by themselves or outsource the recycling to Japan's producer responsibility organization (PRO), the Japan Containers and Packaging Recycling Association. Alternatively, they can be exempt from obligations by collecting waste packaging directly from consumers rather than going through municipalities and recycling the collected items.

continued on next page

Box 7 *continued*

The basic roles of each stakeholder—which has both physical and financial responsibilities—are defined as follows:

(i)   Consumers—responsible for source sorting

(ii) Municipalities—responsible for sorted collection

(iii) Producers—responsible for recycling

(iv) National government—required to endeavor to increase public understanding through educational and publicity activities

The act stipulates that municipalities shall endeavor to take measures necessary to carry out the sorted collection of waste packaging in their areas, but the final decision is left to each municipality; if a municipality agrees with the purpose of the act and formulates a municipal sorted collection plan, the municipality is then required to implement this plan.

Impacts of the policy

The amount of waste ending up in landfills decreased significantly, reducing the number of glass bottles by 39%, PET bottles by 72%, other paper packaging by 60%, and other plastic packaging by 76% after the implementation of the extended producer responsibility-based recycling system. The incineration of PET bottles and other plastic packaging also contributed to this reduction in the quantity of waste landfilled.

The Packaging Recycling Act in Japan had a significant impact on promoting product designs that aimed to reduce waste containers and packaging. It led to the introduction of lighter and thinner products, products with no aluminum lining, flexible packaging, and material changes. Government councils received reports of reductions in various containers and packaging, with reduction rates of several percentage points to more than 60%. Councils also implemented recycling-friendly designs—such as using colored plastic film as labels and adding dotted lines to shrink film for PET bottles—to improve recycling processes and reduce costs.

Estimated Quantities of Four Types of Waste Dumped in Landfills in Japan

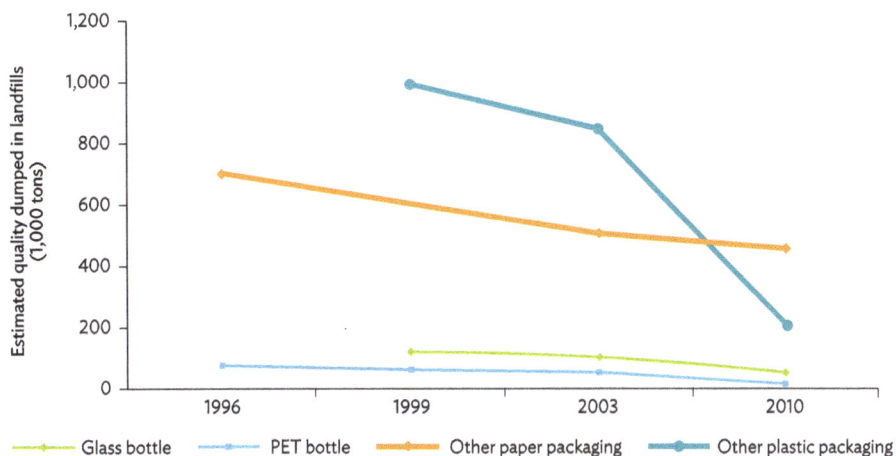

PET = polyethylene terephthalate
Source: OECD. 2013. *The Packaging Recycling Act: The Application of EPR to Packaging Policies in Japan.*

continued on next page

Box 7 *continued*

Key success factors

(i) **Preventing free-riding**. The Government of Japan—particularly the Ministry of Agriculture, Forestry and Fisheries and its Regional Agricultural Administration Offices—takes measures to regulate designated producers and prevent free-riding. These measures include inspections to verify the status of designated producers and penalties for those who do not meet their recycling obligations. The PRO collaborates with the national government by publicizing a list of compliant designated producers and raising awareness through consultations and briefings.

(ii) **Transparency concerning recycling commissions.** Data on parameters used to calculate unit recycling commissions, bid-winning prices, and running costs are submitted annually by the national government to government councils and are disclosed on their website.

(iii) **Transparency on compliance of designated producers**. The PRO maintains a public database containing detailed information about designated producers, including their names and amounts paid for recycling and contributory commissions.

(iv) **Transparency in selecting recyclers.** The PRO conducts competitive bidding for selecting recyclers, with criteria and methods posted on their website. Bidding results—including unit contract prices—are also disclosed.

(v) **Transparency on material flow**. The PRO shares information on waste packaging material flows, including quantities collected, recycled, and sold. This information is available on their website.

(vi) **Transparency on costs**. While up-to-date data on municipal government costs related to waste packaging activities are limited, unit recycling prices, total commission amounts, and financial statements of the PRO are published online.

PET = polyethylene terephthalate.
Source: OECD. 2013. *The Packaging Recycling Act: The Application of EPR to Packaging Policies in Japan.*

4.3 Institutional Arrangements

Institutional arrangements refer to the establishment and organization of various governmental cooperation arrangements aimed at facilitating effective watershed governance and management. Because river distributaries can span multiple local settlements and cross international boundaries, the actions of entities upstream can have a significant impact on downstream areas, underscoring the importance of having the right institutions in place for the effective management of entire river basins. Establishing the appropriate institutional framework also acts as a catalyst for implementing other recommended approaches outlined in this study; the right institutional setup ensures the effective application of NBS, the proper enforcement of various policy measures, and the success of financing mechanisms that support the implementation of these approaches.

Transboundary water management cooperation—both at the intergovernmental and intragovernmental levels—is a critical element in achieving various aspects of sustainable development. Effective water governance, data sharing, conflict resolution mechanisms, and investment in water infrastructure are key components of successful transboundary water management cooperation. Transboundary waters account for 60% of the world's freshwater flows and 153 countries have territory within at least one of the 286 transboundary river and lake basins and

592 transboundary aquifer systems.[195] Yet, only 32 countries have 90% or more of their transboundary basin area covered by operational arrangements, and only 24 countries report that all their transboundary basins are covered by cooperation arrangements.[196]

The remainder of this chapter delves into how institutional arrangements can be developed to combat river pollution in Southeast Asia from both the intergovernmental and the intragovernmental perspectives.

4.3.1 Intergovernmental Cooperation

Overview

The Convention on the Protection and Use of Transboundary Watercourses and International Lakes (Water Convention) is the UN's international legal instrument and intergovernmental platform that aims to ensure the sustainable use of transboundary water resources by facilitating cooperation.[197] When negotiated and implemented equitably and legitimately, transboundary water cooperation arrangements have the potential to help improve water management and cooperation throughout an entire basin, which can result in a large number of direct and indirect economic, social, and environmental benefits for all stakeholders.[198]

There are three major types of institutional arrangements for interstate agreements on transboundary waters: (i) without designation of an institution to implement the agreement, (ii) the appointment of plenipotentiaries (government representatives), and (iii) the establishment of a joint commission responsible for the implementation of the agreement. A few agreements on transboundary waters do not provide for the establishment of any body or institutional mechanism. This approach is typical for agreements that regulate a narrow area of cooperation, e.g., the Agreement between the Government of the Russian Federation and the Government of the People's Republic of China Concerning Guidance of Joint Economic Use of Separate Islands and Surrounding Water Areas in Frontier Rivers (1997). Often countries enter into agreements that do not envisage any bodies or other institutional mechanisms but subsequently realize the need to establish an institutional mechanism to streamline implementation.

Joint commissions prevail in international practice. In general, the establishment of plenipotentiaries is institutionally weaker than that of the joint commissions as plenipotentiaries typically lack additional staff, other organizational structure, and financial resources required for implementing the agreement and decisions taken. Plenipotentiaries often receive public criticism for failing to establish mechanisms to disseminate information or ensure public participation and involvement of stakeholders (e.g., NGOs, youth, women, water user associations, businesses, and local authorities). In contrast, close cooperation with NGOs and their participation in commission work, as well as wide public awareness, have become normal practice for Western European joint commissions.[199]

These agreements that arise from the Water Convention can encompass various scopes, including entire basin coverage; partial basin coverage; boundary waters; or specific projects, programs, or use scenarios of transboundary watercourses. International practice increasingly favors comprehensive watercourse agreements involving all riparian states to facilitate the basin-wide application of integrated water resources management. The functions of these joint bodies are further detailed in the tasks they are entrusted with in the agreements. The Water Convention lists the minimum tasks that joint bodies established under the convention shall be entrusted with. Joint bodies typically possess the following functions:

195 UNWATER. *Transboundary waters.*
196 UN.org. 2023. *Transboundary Water Management Cooperation Crucial for Sustainable Development, Peace, Security.* Speakers Stress at Conference's Fourth Interactive Dialogue; UNWATER. *Transboundary waters.*
197 UNECE. *The Water Convention and the Protocol on Water and Health.*
198 UNECE. 2021. *Practical Guide for the Development of Agreements or Other Arrangements for Transboundary Water Cooperation.*
199 UNECE. 2009. *Capacity for Water Cooperation in Eastern Europe, Caucasus and Central Asia.*

(i) **Coordination and advisory function**, which includes coordination of and assistance to riparian states in their activities to implement the agreement.

(ii) **Executive function**, which includes direct activities of a joint body to implement the agreement.

(iii) **Control of implementation and dispute settlement function**, which includes monitoring of implementation, reporting on implementation, and settling differences and disputes (footnote 199).

Background

The Convention on the Protection and Use of Transboundary Watercourses and International Lakes (Water Convention)—which entered into force in 1996—imposes obligations on parties to prevent, control, and reduce transboundary impacts while safeguarding transboundary waters for ecologically sound management. It encourages equitable water use and ecosystem conservation, guided by principles like the precautionary and polluter-pays principles. The convention also emphasizes pollution prevention at the source, environmental impact assessments, and the best available technology. Riparian parties, or the parties bordering the same transboundary waters, are required to establish joint bodies through bilateral or multilateral agreements, aligning these arrangements with convention principles to foster cooperation among riparian states (footnote 177).

Considerable expertise and best practices have been accumulated through joint commissions in Europe and worldwide regarding the institutional aspects of establishing and operating joint bodies. These include rules of procedure, principles and procedures for decision-making, arrangements for the secretariat, and the regulation of legal personality. Joint commissions have also devised mechanisms to facilitate public participation such as granting observer status, creating working groups for collaboration with NGOs and other stakeholders, and organizing stakeholder conferences, among others.

Implementation

No single existing joint body should be considered as the standard model since these bodies are established with specific waters and tasks, within various political, economic, and social contexts. The implementation of decisions made by a joint body depends on cooperation between the joint body and the national authorities of participating countries. This cooperation can be ensured by appropriate representation in the joint body, including national authorities, ministries, and agencies responsible for water management and protection. Such cooperation is further strengthened by clear reporting mechanisms. Another method to enhance implementation is for participating countries to appoint competent authorities responsible for implementation, or for the joint body to establish additional structures at the national level.

Watercourse agreements that establish joint bodies typically include provisions about financing, with financial commitments varying based on the institutional mechanism and organizational complexity. Generally, each party covers the expenses related to its representatives and experts participating in joint body activities and the costs of monitoring within its territory. For example, under the 1990 Convention on the International Commission for the Protection of the Elbe, each party bears the costs of its representation in the commission, working groups, and territorial monitoring (footnote 199).

Contributions from parties constitute the primary budget source, typically determined by the agreement or later consensus within the joint body. Agreements often specify differentiated contributions based on various criteria (such as share of basin or per capita income), but equal contributions among participating countries are also possible, as seen in the Sava Commission. The Convention on Cooperation for the Protection and Sustainable

Use of the Danube River mandates equal contributions, subject to unanimous decisions by the International Commission for the Protection of the Danube River (ICPDR). However, joint commissions may encounter difficulties with receiving contributions.

Joint commissions can seek further funding from the private sector for specific projects. The ICPDR, in 2005, established the Principles for Cooperation and Relations with Business and Industry, emphasizing that such cooperation should not impede the commission's autonomy. Since then, the ICPDR has partnered with Coca-Cola to raise public awareness and engage in projects aimed at conserving and safeguarding freshwater ecosystems within the Danube River basin. However, the commission should note that financial support from the private sector should be reserved for special projects beyond its core activities (footnote 199).

Applicability to Southeast Asia

The Mekong River Commission (MRC) is one of the most prominent intergovernmental organizations dedicated to managing shared water resources in Southeast Asia. The MRC was established in 1995 by the Agreement on the Cooperation for the Sustainable Development of the Mekong River Basin between the governments of Cambodia, Lao PDR, Thailand, and Viet Nam. Two upstream riparian countries—Myanmar and the PRC—are dialogue partners to the MRC. The MRC produces important data and analysis and works to strengthen norms of basin-wide cooperation, both of which can help to encourage more sustainable approaches to Mekong River development.[200]

The MRC also provides platforms for basin-wide dialogue on contentious issues surrounding Mekong River development and encourages more public participation in Mekong River governance. However, the MRC has weaknesses regarding its efficacy as a platform for Mekong River governance relative to other international joint commissions. For example, even though the consultation processes for two previous mainstream dams—Xayaburi and Don Sahong—ended in disagreement with downstream countries calling for further studies of the social and environmental impacts of each dam, Lao PDR nevertheless moved forward unilaterally with construction. The 1995 Mekong Agreement that created the MRC does not give downstream states veto power over upstream development projects and no legal mechanism exists to punish states that fail to follow through with the MRC's principles or procedures.[201]

Challenges and key success factors

These are the four biggest challenges in the implementation of a joint body:

(i) **Political and diplomatic cooperation**. Differing national interests, geopolitical tensions, and historical conflicts can hinder collaboration on transboundary water management.

(ii) **Resource allocation and funding**. Deciding on a fair and equitable distribution of costs and responsibilities for monitoring, enforcement, and infrastructure development can lead to disputes.

(iii) **Poor implementation**. Poor implementation of a joint body's decisions can arise because of the lack of resources, insufficient motivation among national authorities, inadequate representation of national authorities in the joint body, and the lack of coordination at the national level.

(iv) **Lack of public participation**. Lack of mechanisms for public participation and stakeholder involvement— as well as limited access to information developed by joint bodies—can hinder effective river basin management. Additionally, the absence or improper implementation of provisions for disseminating information can exacerbate these challenges (footnote 199).

---

[200] MRC. *Mekong River Commission*.
[201] G. Neusner. 2016. Why the Mekong River Commission Matters. *The Diplomat*. 7 December.

These are the four main success factors when implementing a joint body:

(i) **Mutual trust**. Mutual trust among the riparian states and the motivation to cooperate are prerequisites for entering into agreements and establishing joint bodies. At the same time, even when such trust does not exist, cooperation may start with joint activities of national authorities on technical issues or in specific areas of cooperation, as well as from joint activities of NGOs and other stakeholders. When a basin-wide agreement by all riparian states cannot be reached, cooperation may start from an agreement and a joint body established by some riparian states to attract all riparian states to such cooperation in the future.

(ii) **Adequate financing**. It is important to ensure the financial sustainability of a joint body by defining the financial commitments of the riparian parties and by analyzing possible additional funding mechanisms.

(iii) **Selecting the right organizational structure**. An organizational structure that allows for developing and adopting decisions as well as implementing them is required. This presumes the existence of decision-making, executive, and working bodies, including a permanent organ to support the activities of a joint body. It also presumes a clear definition of tasks and functions for each element of an organizational structure.

(iv) **Stakeholder involvement**. A joint body should conduct a stakeholder analysis to ensure stakeholder participation in negotiations and to develop mechanisms for stakeholder participation in a joint body's activities (footnote 199).

Box 8 presents an example of institutional arrangements for interstate agreements.

Box 8: The International Commission for the Protection of the Danube River

Background

The International Commission for the Protection of the Danube River (ICPDR) is a transnational body established to implement the Danube River Protection Convention. The ICPDR formally comprises the delegations of all contracting parties to the Danube River Protection Convention but has also established a framework for other organizations to join.

The Danube River Protection Convention forms the overall legal instrument for cooperation on transboundary water management in the Danube River Basin. The convention was signed in June 1994 in Sofia, Bulgaria, and came into force in 1998. It aims to ensure that surface waters and groundwater within the Danube River Basin are managed and used sustainably and equitably.

In 2000, the ICPDR contracting parties nominated the ICPDR as the platform for the implementation of all transboundary aspects of the European Union Water Framework Directive. The successful implementation of the Water Framework Directive is therefore clearly high on the political agendas of the countries of the Danube River Basin District. In 2007, the ICPDR also took responsibility for coordinating the implementation of the European Union Floods Directive within the Danube River Basin.

continued on next page

Box 8 *continued*

Three Goals of the International Commission for the Protection of the Danube River

(i)   A Cleaner Danube —this means reducing pollution from settlements, industry, and agriculture.

(ii)   A Healthier Danube—this means protecting rivers as ecosystems that provide a living environment for aquatic animals and plants, as well as services for people such as drinking water and recreation.

(iii) A Safer Danube—this means a safer environment for people to live without the fear of major flood damage.

How the International Commission for the Protection of the Danube River Is Run

(i)   **Setup**. The ICPDR is a small and lean organization. Its organizational set up consists of the commission, bringing together member state representatives as a conference of parties (in charge of decision-making); expert groups and task groups consisting of national technical experts developing recommendations on specific issue matters; and the secretariat of the commission.

(ii)   **Decision-making**. Decision-making is primarily based on consensus, although legal majority decisions by 80% of delegations are possible. In practice, majority-based decision-making is rarely used. Instead, expert groups serve as forums for discussing issues and creating proposals for solutions that all member countries must accept before reaching the formal decision-making level.

(iii) **Financing**. ICPDR's funding mainly comes from its member states, and its overall financial needs are relatively low, with the Secretariat having an annual budget of about €1 million. Member states cover their costs for participating in the cooperation process and implement agreed-upon activities themselves, as the organization strongly emphasizes coordination. All member states have agreed on an equal cost-sharing arrangement for common expenses. In 1999, member states established temporary exceptions from equal cost-sharing, grouping according to their economic capacity and thus their ability to financially contribute to the river basin organization. This allowed weaker member states a temporary reduction in contributions and committed economically more advanced states to temporarily cover these shares.

(iv)   **Dispute resolution**. The ICPDR plays a formal role in dispute resolution, initially facilitating negotiations among disputing parties. If disputes cannot be resolved internally, they can be referred to external third-party resolution, such as the International Court of Justice or an arbitration tribunal. However, historical disagreements are typically resolved diplomatically, with the ICPDR primarily focused on preventing and mitigating disputes. Regional integration among member countries contributes to the infrequent escalation of disagreements.

continued on next page

Box 8 continued

Organizational Chart of the International Commission for the Protection of the Danube River

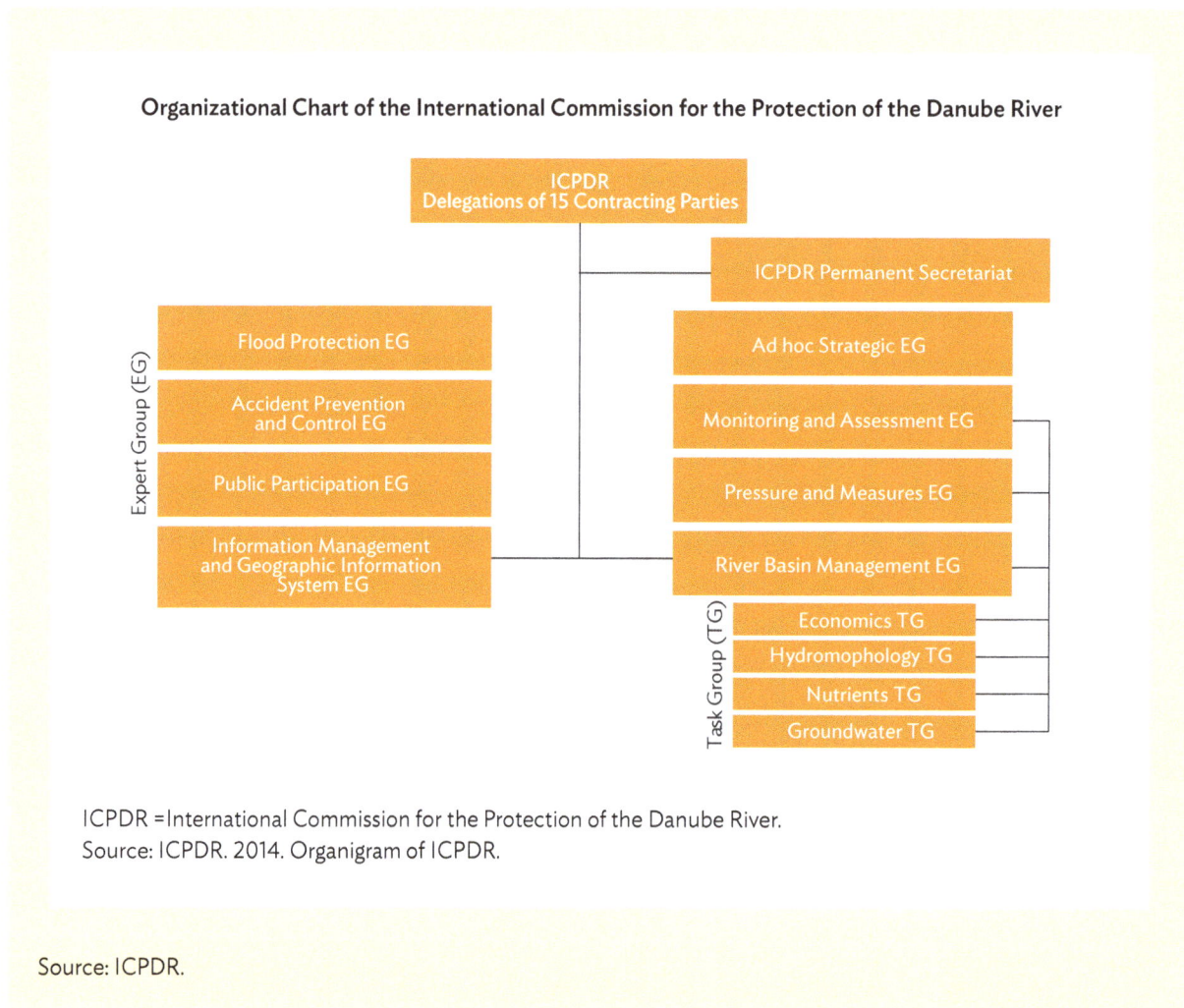

ICPDR = International Commission for the Protection of the Danube River.
Source: ICPDR. 2014. Organigram of ICPDR.

Source: ICPDR.

4.3.2 Intragovernmental Cooperation

Water governance involves multiple actors across different institutional levels. Many of the prevailing and impending water crises are not necessarily because of water scarcity or a lack of technological expertise but rather stem from poor management and governance of available water resources and infrastructure. Within a country, various actors—both public and private, at different institutional levels—contribute to water governance. Typically, municipalities are delegated the responsibility for providing water services in cities, while regional, national, and private entities may play vital roles in terms of oversight, technical knowledge, and funding.[202]

Several complexities and obstacles contributing to coordination difficulties within an intragovernmental river pollution institutional framework include the following:

---

[202] E. Lieberherr and K. Ingold. 2019. Actors in Water Governance: Barriers and Bridges for Coordination. *Water.* 11(2). p. 326.

(i)   **Fragmented responsibilities**. Fragmentation frequently arises in regions where multiple administrative areas share hydrological connectivity or fall under the jurisdiction of various governing authorities with interconnected responsibilities. In terms of water pollution control, interjurisdictional fragmentation arises when several local governments claim equal rights to the limited assimilative capacity of a shared body of water.[203]

(ii) **Limited resources**. Inadequate resources within water management agencies can significantly hinder their capacity for effective planning and, consequently, their ability to coordinate effectively. When operating in resource-constrained contexts, these agencies often contend with competing priorities for limited resource allocation. In such scenarios, agencies may prioritize their immediate mandates over pollution management, thereby undermining coordination efforts aimed at addressing common environmental challenges.[204]

(iii) **Political interference**. Decisions regarding water management and pollution control may be influenced by political considerations rather than sound environmental practices. Political influence can pressure the decision-making process, leading to suboptimal outcomes. This interference may take various forms, such as prioritizing short-term political goals over long-term environmental sustainability.[205]

Poor coordination among various actors in a country can lead to significant problems in managing river pollution:

(i)   **Insufficient enforcement**. Insufficient enforcement can result in inconsistent enforcement practices, allowing some polluters to escape penalties, undermining the credibility of pollution control efforts, and potentially exacerbating environmental harm.[206]

(ii) **Regulatory gaps**. Regulatory gaps can emerge because of poor coordination among water management agencies, leading to inconsistencies and deficiencies in pollution control standards and regulations. These gaps can create legal ambiguities, hinder comprehensive pollution management, and leave room for polluters to exploit regulatory weaknesses.[207]

(iii) **Inefficient resource allocation**. When water agencies do not coordinate effectively, resources such as funding, personnel, and equipment may be allocated inefficiently. This can lead to overlapping efforts in some areas while critical pollution hotspots remain unaddressed, resulting in wasted resources and suboptimal pollution control.[208]

(iv) **Lack of data sharing and monitoring**. Effective pollution management relies on accurate data collection, sharing, and analysis. Poor coordination among water agencies may impede the sharing of critical pollution data and hinder the development of comprehensive monitoring programs, limiting the ability to track pollution sources, assess trends, and make informed decisions for pollution control.[209]

Box 9 elaborates about the PRC's River Chief System (He-Zhang) and Box 10 elaborates about France's 6 Water Agencies (Les 6 agences de l'eau). These cases exemplify the establishment of systems within a national water management framework, facilitating efficient intragovernmental cooperation to achieve optimal results in addressing river pollution.

---

203 H. Liu et al. 2019. The River Chief System and River Pollution Control in China: A Case Study of Foshan. *Water.* 11(8). p. 1606.

204 Z. Zhao et al. 2020. Problems and Countermeasures of River Management in the Process of Rapid Urbanization in China. *Water.* 12(8).

205 UNESCAP. 2019. *Tackling Water Pollution and Promoting Efficient Water Use in Industries.*

206 A. Heyes. 2000. *Implementing Environmental Regulation: Enforcement and Compliance.* OECD.

207 Y. Chu and J. Wang. 2019. The Dilemma and Solution of Water Pollution Control from the Perspective of Environmental Regulation. *E3S Web of Conferences.* 136.

208 OECD. 2015. *Water Resources Allocation: Policy Highlights.*

209 S. Gallaher and T. Heikkila. 2014. Challenges and Opportunities for Collecting and Sharing Data on Water Governance Institutions. *Journal of Contemporary Water Research and Education.* 153(1). pp. 66–78.

Box 9: People's Republic of China's River Chief System (He-Zhang)

Overview

In December 2016, the government created a new system of river chiefs, the River Chief System (RCS), for the country's waterways. A revision of the 2008 Water Pollution Prevention and Control Law—scheduled to take effect in 2018—codifies the responsibility of river chiefs to supervise water quality, enforce pollution regulations, and oversee ecological restoration efforts.

This system names a single person—typically a senior official at the local, county, or provincial level—to be responsible for each stretch or section of every major lake and waterway. These officials are responsible for meeting environmental protection and water-quality targets in their jurisdictions. River chiefs at the provincial level are also responsible for dealing with interjurisdictional issues.

The river and lake chief system effectively makes the leaders of each province, city, county, and township responsible for core water management functions, supported by a dedicated office at the county level and above.

The creation of these positions reflects the fact that these policy priorities have often been hindered by interjurisdictional and intergovernmental coordination problems. The river and lake chiefs are expected to ensure that officials of various departments under their control work together to achieve key water policy objectives

Structure and Concrete Operational Process of the River Chief System at the County Level in the People's Republic of China

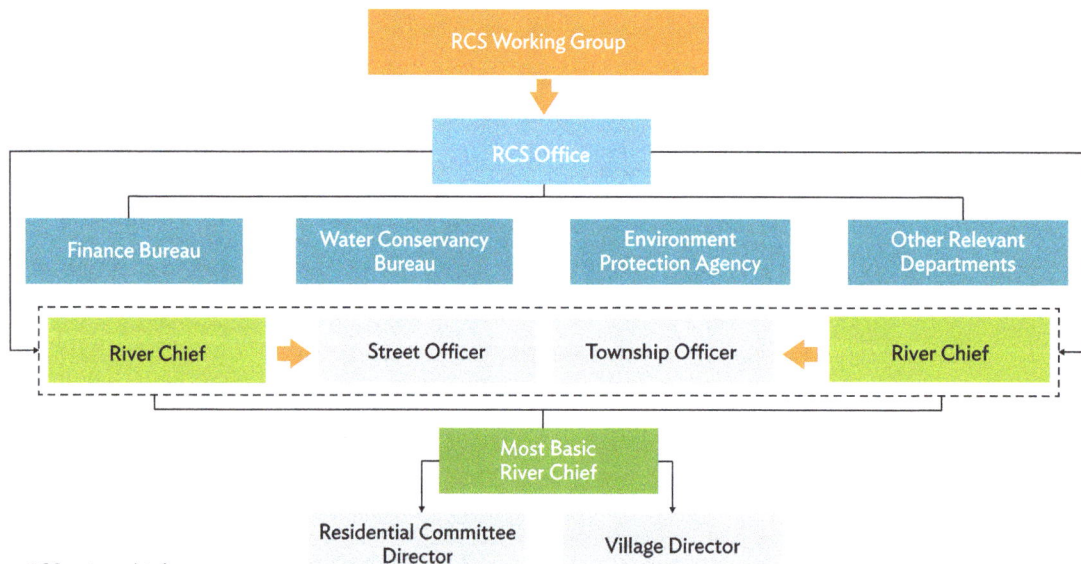

RCS = river chief system.
Source: B. Wang et al. 2021. River Chief System: An Institutional Analysis to Address Watershed Governance in China. *Water Policy*. 23(6). pp. 1435–1444.

continued on next page

Box 9 *continued*

Benefits of the system

The RCS provides three key advantages:

(i)   The RCS serves to **define the precise roles and responsibilities of river basin management** entities. It establishes a structured approach to water environment management akin to administrative or official contracting. This system provides a clear delineation of responsibilities, with government departments and officials at various levels being held accountable for water environment management within their respective river basins.

(ii) The RCS **enhances the flexibility of collaboration among various government departments** involved in river basin water environment management. Local government leaders actively engage in managing water environment affairs within their jurisdictions, with RCS offices serving as coordinating bodies responsible for tasks such as supervision, guidance, inspection, and communication. Importantly, the RCS does not replace existing water-related departments but rather complements their functions.

(iii) The RCS **employs a one-vote environmental accountability mechanism** to emphasize the significance of watershed water environmental governance. This approach encourages local government officials to consider environmental management alongside economic development. The assessment method for river basin environmental governance places greater political responsibility and pressure on local government leaders, motivating them to prioritize environmental governance. Additionally, the results of natural resources and environmental management during their tenure are crucial performance evaluation criteria, leading to the implementation of a lifelong accountability system for ecological and environmental harm.

Sources: World Bank Group. 2019. *Vietnam: Toward a Safe, Clean, and Resilient Water System.* B. Wang et al. 2021. River Chief System: An Institutional Analysis to Address Watershed Governance in China. *Water Policy.* 23(6). pp. 1435–1444.

Box 10: France's 6 Water Agencies (Les 6 agences de l'eau)

Overview

In 1964, the French defined a water management framework which resulted in the creation of six water agencies that are responsible for managing and conserving water resources and aquatic environments. They operate under the authority of the Ministry for Ecological Transition and the Ministry of Economy, Finances, and Recovery. An example of such an agency is the *Agence de l'Eau Seine-Normandie,* which oversees the Seine-Normandy basin. The agency funds investment projects and initiatives aimed at preserving water resources and combating pollution, with funding derived from fees collected from all users. These collected revenues are then redistributed in the form of subsidies and/or advance payments to local authorities, economic operators, farmers, or associations engaged in activities aimed at protecting the natural environment. Additionally, these agencies also provide technical assistance for water-related general interest measures, ensuring the sound and cost-effective management of water resources and aquatic environments.

The water policy implemented by the water agencies has had a positive environmental impact. As of 2023, 41% of natural waters have achieved a water quality status of "good" or "very good." The French water agencies have also been internationally recognized for providing an effective water management structure. In 2000, the European Union created its river basin district model based on the French model.

continued on next page

Box 10 *continued*

Benefits of the system

The water agency system provides seven key advantages:

(i) **Stakeholder engagement.** They engage various stakeholders, including agriculture, industry, nongovernment organizations, civil society, and scientists, for collaborative water management. They also form partnerships with local and international entities, such as the National Forestry Office and foreign countries like Bulgaria and Viet Nam, to develop solutions for water management.

(ii) **Political commitment.** Multiple government ministries are involved, showing a strong political commitment to water policy and environmental protection.

(iii) **Public confidence.** They hold public consultations to gauge public opinions, ensuring alignment with public interests.

(iv) **Clarity of objectives.** Objectives are clear and measurable, focusing on water quality, pollution control, sustainability, and public awareness.

(v) **Strength of evidence.** The agencies have a strong legal framework and historical expertise in basin management, making it easier to meet water quality standards.

(vi) **Feasibility.** Their objectives are financially, legally, and technically feasible, supported by substantial funding sources. They have an annual income of €1.8 billion from water user bill payments in addition to government funding.

(vii) **Management and measurement.** Water management is decentralized, involving elected governments, and they use a national Water Information System for effective data collection, monitoring, and reporting.

Sources: EU. 2021. Ensuring that Polluters Pay: France; Centre for Public Impact. 2016. The French water management agencies.

4.4 Financing Mechanisms

A concept conceived in 2012, the blue economy refers to the sustainable use of ocean resources for economic growth, improved livelihoods and jobs, and ocean ecosystem health.[210] The blue economy combines innovative marine activities like aquaculture, biological sciences, offshore renewable energy, and bioprospecting with traditional marine-based industries like fishing, maritime transport, and tourism in the rivers, seas, and oceans. Yet, these industries are still gravely underfunded.

Southeast Asia faces a large infrastructure financing gap in urgent need of attention to ensure sustainable economic growth. ADB estimated ASEAN's total infrastructure investment requirement at $2.8 trillion (baseline estimate) and $3.1 trillion (climate-adjusted estimate), placing the annual investment need at $184 billion to reduce poverty and $210 billion to respond to climate change.[211] Investment needs vary considerably across the blue economy focal areas and market segments. Of those segments with the highest financing needs in Southeast Asia, resilient ports tops the list, as it faces existential threats from rising seas and storms. Marine offshore wind renewable energy comes second, and the financing gap in this market segment represents the investment needed to realize the full potential of this technology in countries in the region, in line with what is being done in Europe and the PRC.

---

[210] World Bank. 2017. *What Is the Blue Economy?*

[211] ADB. 2023. *Innovative Financing Can Help Bridge Southeast Asia's Infrastructure Financing Gap.*

Investments required for pollution control such as solid waste management, wastewater management, and nonpoint source pollution management have also been identified to be facing significant financing gaps (Figure 20).

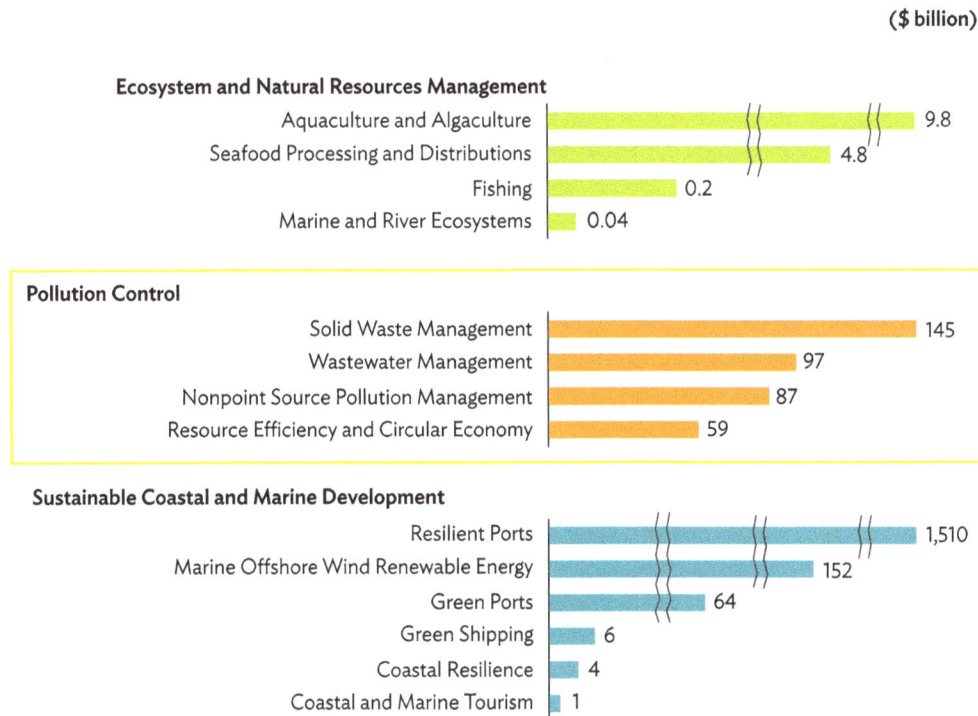

Figure 20: Finance Gap in Southeast Asia: Investments Needed by 2030 to Meet Sustainable Development Goals

($ billion)

Ecosystem and Natural Resources Management
- Aquaculture and Algaculture — 9.8
- Seafood Processing and Distributions — 4.8
- Fishing — 0.2
- Marine and River Ecosystems — 0.04

Pollution Control
- Solid Waste Management — 145
- Wastewater Management — 97
- Nonpoint Source Pollution Management — 87
- Resource Efficiency and Circular Economy — 59

Sustainable Coastal and Marine Development
- Resilient Ports — 1,510
- Marine Offshore Wind Renewable Energy — 152
- Green Ports — 64
- Green Shipping — 6
- Coastal Resilience — 4
- Coastal and Marine Tourism — 1

Source: ADB. 2022. *Financing the Blue Economy.*

While the blue economy offers numerous opportunities for economic growth and environmental sustainability, several significant obstacles contribute to a financing gap for its advancement. In Asia, the public sector has been responsible for as much as 92% of total infrastructure financing, yet this is still insufficient to bridge the widening deficit between funding needs and actual capital spending.[212] Similar to blue economy projects, they require substantial upfront investments with uncertain payback periods, which deter private and public investors. Also, the blue economy often operates in complex regulatory environments with overlapping jurisdictions. Inconsistent or unclear regulations, lack of enforcement, and governance challenges create further uncertainties for investors.

Financing tools for addressing water pollution in this region are predominantly conventional, relying on government budgets, international aid, and traditional development loans. Some of the financing options include subsidies, microfinance, partnerships, conservation trust funds, output-based aid, and sustainability-linked loans that support water initiatives aimed at prevention, mitigation, and remediation. However, there is a notable lack of comprehensive case studies on innovative financing tools specifically tailored to combat river water pollution in the region.

---

[212] ADB. 2018. *Closing the Financing Gap in Asian Infrastructure.*

This chapter focuses on three key financing mechanisms:

(i)   blue bonds,

(ii)   public–private partnerships (PPPs), and

(iii)   blended finance.

By deploying these financing mechanisms, Southeast Asian countries can potentially mobilize the resources needed to effectively tackle river water pollution. While the context may differ from the broader water sector, drawing upon these established financing models offers a promising avenue for addressing a critical environmental challenge in the region.

4.4.1   Blue Bonds

Overview

A blue bond is a relatively new form of a sustainability bond. They are a type of debt instrument issued by governments, development banks, or others to raise capital from impact investors to finance marine- and ocean-based projects that have positive environmental, economic, and climate benefits. The blue bond is inspired by the more common green bond concept.[213]

To repay the capital and interest, the issuer of a blue bond must have a source of revenue. ADB supports a model wherein blue bonds fund projects that generate financial returns, otherwise known as a "use of proceeds" bond. ADB's first blue bond—issued in September 2021—is an example of this model. The Bank of China and Nordic Investment Bank have also issued blue bonds with the use of the proceeds model. Other types of blue bonds exist, for example, the Republic of the Seychelles issued a blue bond that—among other revenue streams—used a debt swap to support issuance.

While blue bonds are normally linked with ocean conservation and given the parallels and synergies between water and ocean financing, interviews with experts and other studies suggest that blue bonds can be used to solve water pollution and can also be accessed to fund projects for well-defined sustainable blue economy sectors such as sustainable fishing, ecotourism, waste management sector, and marine renewable energy (Figure 21).[214]

Background

The term "blue economy" was first introduced at the Rio+20 United Nations Conference on Sustainable Development in 2012. Many coastal nations questioned the viability of the green economy in their situations during the Rio+20 preparation phase and pushed for the growing significance of a blue economy strategy.[215] There was substantial discussion about whether the blue bond designation should have been adopted in the first place, as many classic green investors found that simply having the "green" label satisfied their needs. However, there has been an increase in support for the addition of more specific labels, such as orange bonds for gender projects, as well as the explicit blue label.

The value of funding a sustainable blue economy has become increasingly apparent and blue bonds can play a significant role in supporting the growth and sustainability of the blue economy by providing financing mechanisms.[216] The blue economy has a huge funding shortage that can be filled by creating an appeal to investors

---

[213]   World Bank. 2018. *Sovereign Blue Bond Issuance: Frequently Asked Questions.*

[214]   ADB. 2021. *Sovereign Blue Bonds: Quick Start Guide.*

[215]   SPF et al. 2023. *Financing the Blue Economy in Asia.*

[216]   N. Roth et al. 2019. *Blue Bonds: Financing Resilience of Coastal Ecosystems.*

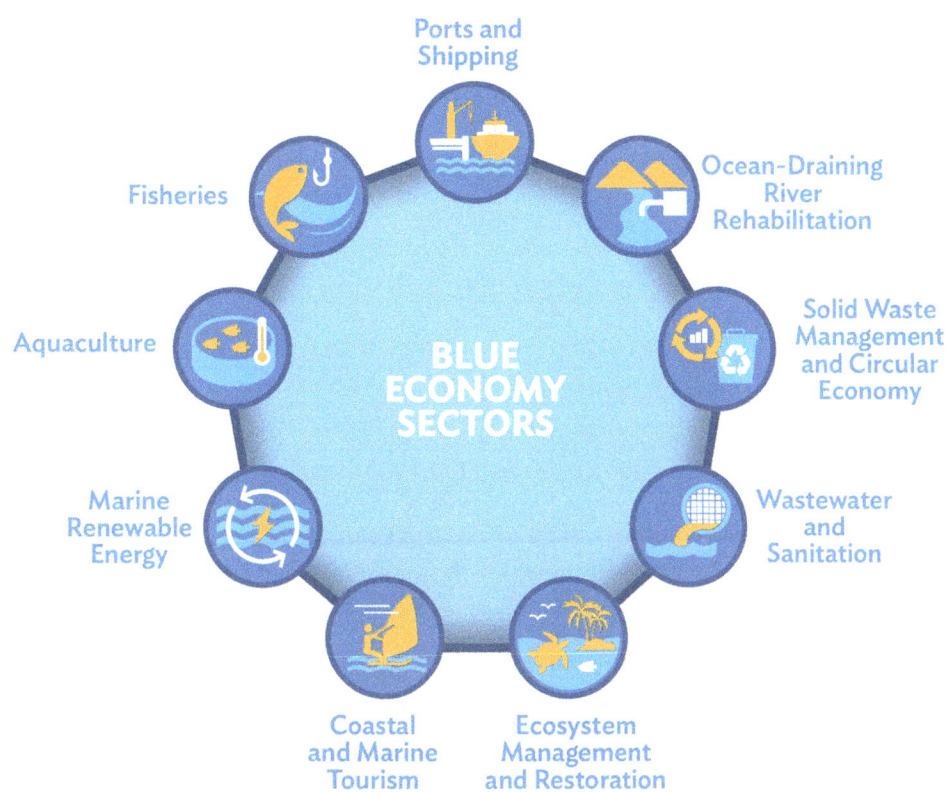

Figure 21: Projects that Can Be Funded by a Blue Bond

Ports and Shipping

Fisheries

Ocean-Draining River Rehabilitation

Aquaculture

Solid Waste Management and Circular Economy

BLUE ECONOMY SECTORS

Marine Renewable Energy

Wastewater and Sanitation

Coastal and Marine Tourism

Ecosystem Management and Restoration

Source: ADB. 2021. *Sovereign Blue Bonds: Quick Start Guide.*

with blue bonds.[217] Investments can create new prospects for growth and development because of the enormous economic potential of the blue economy.[218]

Implementation

Funding for blue economy activities has been raised using a range of designated blue debt financing mechanisms or structures. These include bonds and loans with blue use of proceeds, sustainability-linked bonds, loans with blue performance targets, as well as cutting-edge structures like sovereign debt swaps for climate change adaptation.[219]

Blue bonds can be issued by governments, banks, or corporations. They adhere to the International Capital Market Association (ICMA) Green Bond Principles requirements, including the use of proceeds for projects that address important environmental issues, the establishment of a process for project evaluation and selection, and the use of a formal internal process to monitor the application and management of proceeds and the requirement for annual reporting on the use of proceeds. Figure 22 illustrates the concept of a blue bond.

[217] R. Tirumala and P. Tiwari. 2022. Innovative Financing Mechanism for Blue Economy Projects. *Marine Policy*. 139.
[218] M. Konar and H. Ding. 2022. *A Sustainable Ocean Economy for 2050*. Ocean Economy.
[219] SPF et al. 2023. *Financing the Blue Economy in Asia*.

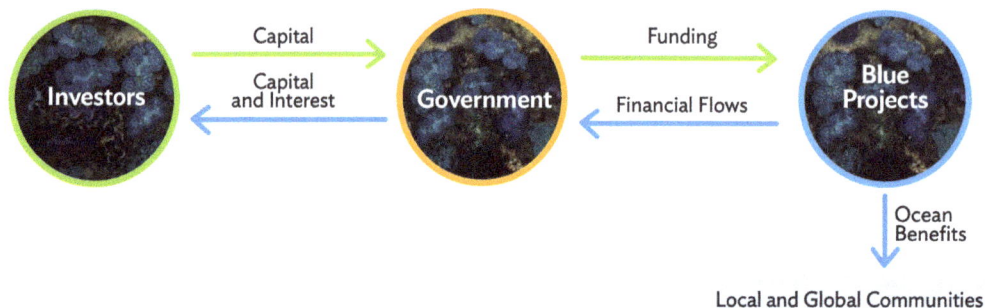

Figure 22: Concept of a Blue Bond

Source: ADB. 2021. *Sovereign Blue Bonds: Quick Start Guide.*

Most bonds invest earnings to solve issues relating to ocean health, including solid waste and wastewater management. For example, some bonds fund "waste disposal facilities at ports and terminals" and "wastewater treatment and water pollution prevention, to reduce discharges into water (mainly phosphorus, nitrogen, organic matter, heavy metals, plastics, and pharmaceuticals)." To maintain the biodiversity of the ocean, other bonds concentrate on safeguarding species and habitats. Protecting biodiversity in "wetlands, rivers and lakes, coastal areas, and open sea zones" are some of the projects that fall under this category (Figure 23).

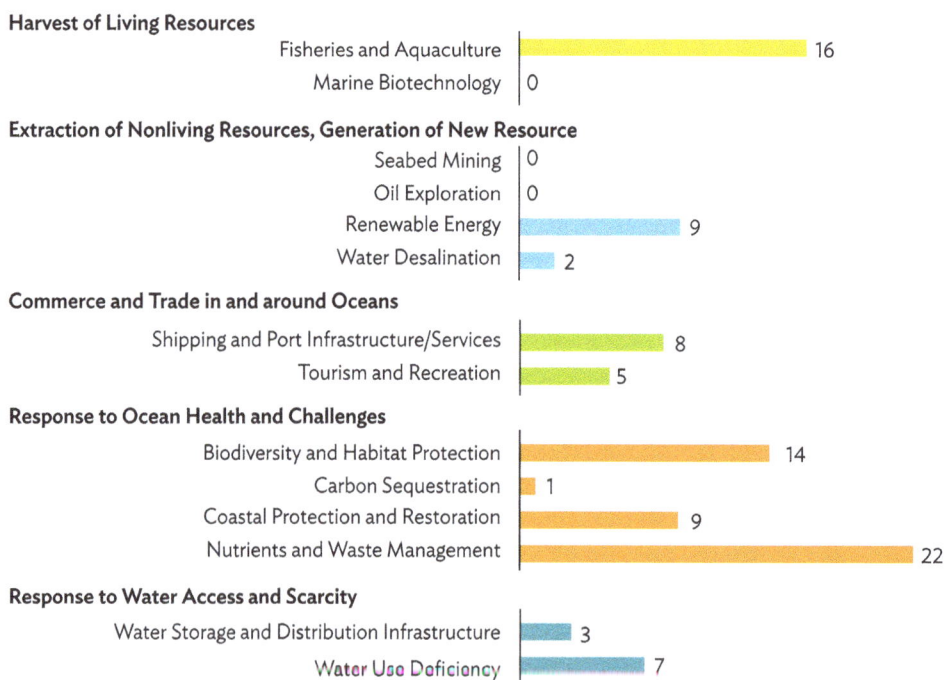

Figure 23: Mapping of Blue Bond Use of Proceeds to the Impact-Area Taxonomy (2018–2022)

Harvest of Living Resources
- Fisheries and Aquaculture: 16
- Marine Biotechnology: 0

Extraction of Nonliving Resources, Generation of New Resource
- Seabed Mining: 0
- Oil Exploration: 0
- Renewable Energy: 9
- Water Desalination: 2

Commerce and Trade in and around Oceans
- Shipping and Port Infrastructure/Services: 8
- Tourism and Recreation: 5

Response to Ocean Health and Challenges
- Biodiversity and Habitat Protection: 14
- Carbon Sequestration: 1
- Coastal Protection and Restoration: 9
- Nutrients and Waste Management: 22

Response to Water Access and Scarcity
- Water Storage and Distribution Infrastructure: 3
- Water Use Deficiency: 7

Source: P. Bosmans and F. Mariz. 2023. The Blue Bond Market: A Catalyst for Ocean and Water Financing. *Journal of Risk and Financial Management.* 16(3). p. 184.

According to a study by Tameo Impact Fund Solutions and Sasakawa Peace Foundation , there have been 51 global blue issuances totaling $9.7 billion since 2018 (Figure 24). Of these, 22 issuances totaling $5.1 billion or 53%) were in Asia.[220] Blue bonds are most appropriate for countries with robust ocean governance, sustainable economic activities, and sizable pipelines of loan projects. Because of the costs involved in issuing a bond, most blue bonds need to be at least $50 million–$500 million in scale (Figure 25) (footnote 214).

Figure 24: Global Blue Bond Issuances

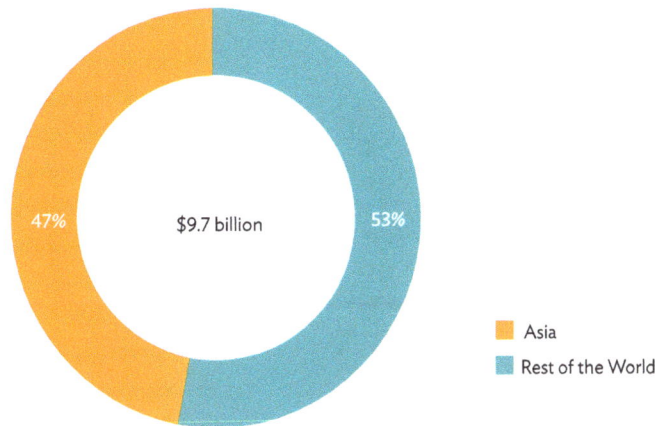

Source: Tameo Impact Fund Solutions and Sasakawa Peace Foundation. 2023. *Financing the Blue Economy in Asia.*

Figure 25: Annual Blue Bond Issuances

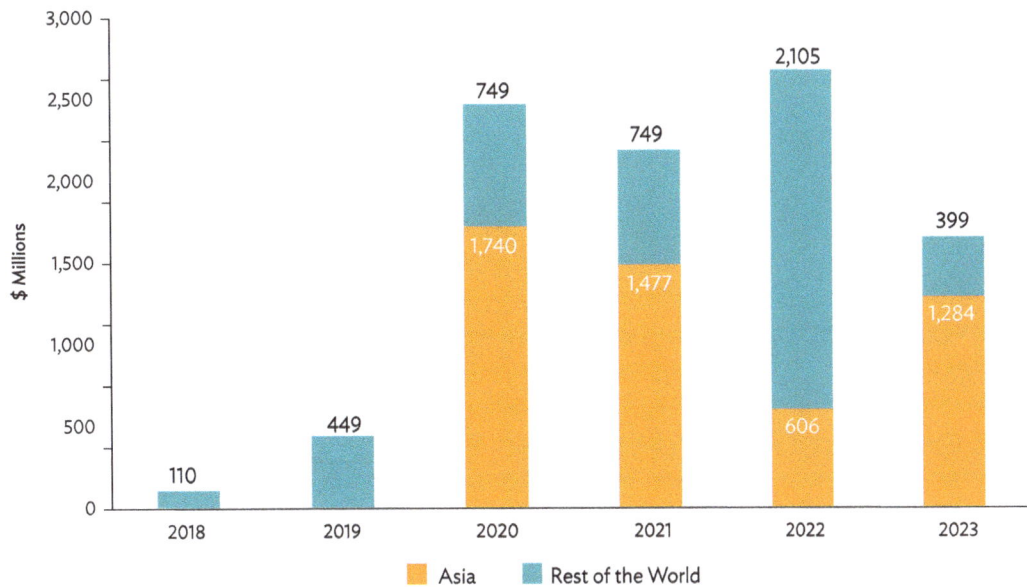

Source: Tameo Impact Fund Solutions and Sasakawa Peace Foundation. 2023. *Financing the Blue Economy in Asia.*

---

220   Tameo Impact Fund Solutions and Sasakawa Peace Foundation. 2023. *Financing the Blue Economy in Asia.*

In 2021, the cumulative total value of blue bonds issued peaked at $1.5 billion, and the market for blue bonds is anticipated to keep expanding in the coming years. Some of the blue bonds issued globally include the following:

(i) In September 2021, ADB issued its first dual-tranche blue bonds denominated in Australian and New Zealand dollars totaling $302 million to finance ocean-related projects in Asia and the Pacific. ADB issued the bonds under its expanded Green and Blue Bond Framework.[221]

(ii) The Korea Eximbank issued a 10-year blue bond valued at $1 billion in January 2023. The bond is offered to raise money for investments in environmentally friendly marine initiatives like the development of marine renewable energy sources and environmentally friendly shipbuilding.[222]

(iii) *Desarrollos Hidráulicos de Cancun* issued Mexico's first blue bond in January 2023 worth $74 million. The funds will be used to finance or refinance initiatives that support SDG 6 and SDG 9, to build resilient infrastructure, support inclusive and sustainable industrialization, and foster innovation. *Desarrollos Hidráulicos de Cancun* created the Blue Bonds Program intending to provide drinking water to the residents of the municipalities where it operates.[223] To further advance the blue taxonomy goals, in January 2023, the Central American Bank for Economic Integration issued its second blue bond on the Japanese market for $53 million.[224]

Applicability to Southeast Asia

The first blue bond in Southeast Asia was issued in the Philippines in 2022.[225] The International Finance Corporation (IFC)—the private investment arm of the World Bank Group—committed to investing $100 million into Banco de Oro Unibank Inc.'s Blue Bond, proceeds of which will be used to help tackle marine pollution and preserve clean water resources while supporting the country's climate goals.

In May 2022, the IFC announced that it will subscribe up to $50 million in a blue bond issued by TMBThanachart Bank (TTB) to help increase access to financing for climate-smart solutions and blue economy projects in Thailand. This is the first blue bond issued by a commercial bank in Thailand, with the IFC helping TTB to structure the bond and assist in developing a blue finance strategy and framework.[226]

In May 2023, Indonesia issued a first-of-its-kind blue bond in the Japanese debt capital market, raising ¥20.7 billion ($150 million). This issuance marks the world's first publicly offered sovereign blue bond aligned with ICMA principles with assistance from the United Nations Development Programme (UNDP), HSBC Bank, and Credit Agricole. Its issuance demonstrates Indonesia's commitment to tap innovative sources of financing for investments that benefit communities and the sustainable use of marine ecosystems.[227]

Challenges

While blue bonds can be an effective financing tool for reducing water pollution, there are several key challenges with their implementation.

---

[221] ADB. 2021. *ADB Issues First Blue Bond for Ocean Investments. News release.* 10 September.

[222] Y. Yoon. 2023. Korea Eximbank Issues Foreign Currency Bonds Worth US$3.5bn. *Business Korea.* 6 January.

[223] BBVA. 2023. *BBVA assisted Desarrollos Hidráulicos de Cancún in the issuance of the first blue bond in Mexico and the first of its kind for the bank.* 19 January.

[224] CABEI. 2023. *CABEI issues its second blue bond less than a month after publishing its Blue Taxonomy.* 1 January.

[225] World Bank. 2021. *Market Study for the Philippines: Plastics Circularity Opportunities and Barriers.*

[226] C. Santiago. 2022. IFC powers ttb's EV-focused first green bond. *The Assest.* 19 October.

[227] UNDP. 2023. *Indonesia Launches the World's First Publicly Offered Sovereign Blue Bond—with UNDP's Support.* 25 June.

(i) **The applicability of blue bonds on river pollution reduction and control is yet to be proven.** Blue bonds are primarily associated with financing marine and ocean-related projects such as marine conservation, sustainable fisheries, and ocean health initiatives, and their application to river pollution control is less established. The applicability of blue bonds to river pollution control can be further explored through the identification of possible revenue sources for the blue bond issuer. Under ADB's Green and Blue Bond Framework introduced in 2021, investments that contribute to pollution control for marine and coastal environments—including the rivers that drain to the ocean (such as management of solid waste, non point source pollution, and wastewater)—could be projects eligible for blue bond issuance.[228]

(ii) **Bond issues are complex, requiring multiple technical experts as well as multiple parties.** These complexities stem from the need to ensure the transparency, credibility, and effectiveness of blue bond projects, which are designed to support marine and ocean-related initiatives. Identifying and evaluating suitable marine and ocean projects that align with the bond's objectives can be a complex process. Expertise in marine biology, environmental science, and project management is often required to assess the potential impact and risks of these initiatives. It can be challenging to ensure blue bonds adhere to regulations and standards, including those related to environmental impact assessments, permits, and legal compliance. Multiple stakeholders— including local communities, environmental organizations, and government agencies—play a crucial role in the success of blue bond projects, and it is often complex to manage these relationships.[229]

(iii) **The lack of standardized definitions, metrics, and expertise by issuers and investors are significant barriers.** There is no universally accepted definition of what constitutes a "blue" project. This lack of standardization can lead to ambiguity and differing interpretations by issuers and investors. Also, blue bonds lack standardized metrics for assessing the positive outcomes of marine and ocean initiatives. Issuers of blue bonds may lack the expertise to assess and select appropriate projects while investors may lack expertise to evaluate the environmental, social, and financial risks and benefits of blue bond investments. Resolving these barriers is crucial to attracting corporations and ensuring the continued growth of the blue bond market.[230]

Box 11 presents the example of the Nordic Baltic Blue Bonds.

Box 11: Nordic Baltic Blue Bond Issuance

Background

Although water resources are abundant in the Nordic and Baltic regions, the pressure on water habitats and local marine and coastal waters has been growing because of human activities and pollution. Thus, the Nordic Investment Bank (NIB) released the first Nordic-Baltic Blue Bond of $220 million in 2019 to fund initiatives that would attempt to alleviate these pressures. Because of the high levels of nitrogen and phosphate release that result in excessive plant and algae development, the Baltic Sea has been particularly impacted by eutrophication, thus requiring investments in wastewater treatment to ensure sustainable urban expansion. NIB then launched its second Nordic–Baltic Blue Bond under the bank's Environmental Bond Framework in 2020.

continued on next page

---

228 ADB. 2021. *Green and Blue Bonds Framework*.
229 ADB. 2019. *Blue Bonds: Financing Resilience of Coastal Ecosystems*.
230 P. Bosmans and F. Mariz. 2023. The Blue Bond Market: A Catalyst for Ocean and Water Financing. *Journal of Risk and Financial Management*. 16(3). p. 184.

Box 11 *continued*

By the end of 2021, NIB had issued €5.8 billion in environmental bonds, including €335 million in blue bonds, which financed 13 water treatment projects that contribute to the reduction of pollution and water-related climate adaptation.

Nordic Investment Bank Environmental Bond Issuance (2011–2021), Including Blue Bonds

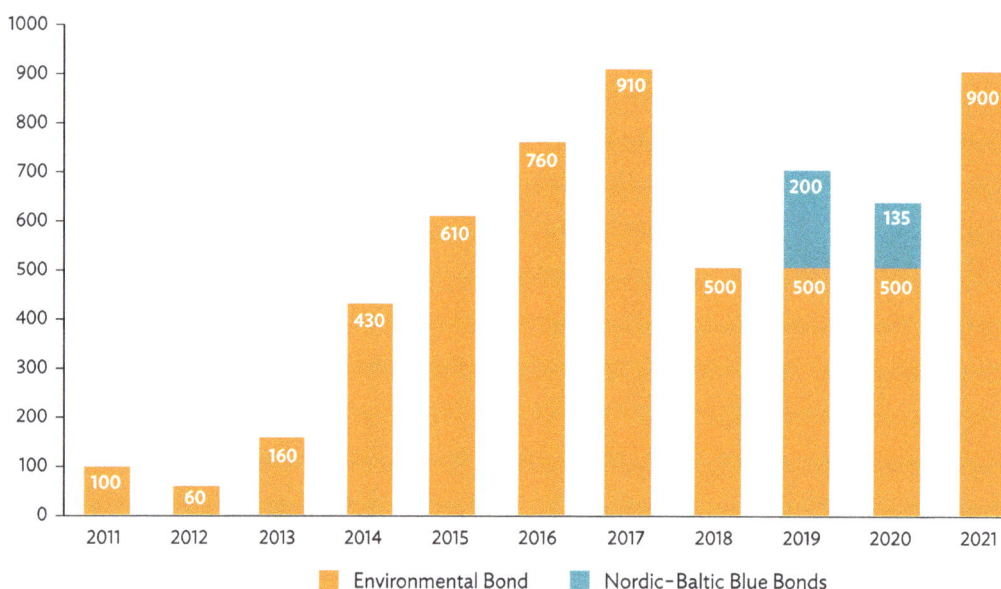

Source: NIB. 2021. NIB Environmental Bond Report.

Impact

The proceeds from the Nordic–Baltic Blue Bond were to be used for the following:

(i)   **Wastewater treatment and water pollution prevention**. Aims to decrease discharges into water (mostly phosphorus, nitrogen, organic matter, heavy metals, plastics, and pharmaceuticals).

(ii) **Stormwater systems and flood protection**. Aids in the reduction of pollution and the acceleration of climate change dependable infrastructure.

(iii) **Protection of water resources**. Reducing groundwater exploitation and contamination while enhancing aquifer recharge.

(iv) **Protection and restoration of water and marine ecosystems**. Projects aiming at the expansion of protected areas, the protection and restoration of water and marine ecosystems, and biodiversity (such as wetlands, rivers and lakes, coastal regions, and open sea zones).

continued on next page

Box 11 *continued*

Implementation

The NIB's idea of "blue" does not include oceans, but rather problems involving water. The bank promotes loans to initiatives addressing wastewater management, the defense of water systems, the conservation of water resources, and the preservation of marine ecosystems and biodiversity. Additionally, it supports a project for disaster mitigation that aims to strengthen flood resilience and robustness against sea-level rises and adverse weather conditions. This example demonstrates how blue bonds can be understood independently of the ocean economy and can be applied to other water-related issues.

Sources: NIB. 2019. *NIB issues first Nordic–Baltic Blue Bond*; NIB. 2020. *NIB launches five-year SEK 1.5 billion Nordic–Baltic Blue Bond*; NIB. 2021. *NIB Environmental Bond Report*; P. Bosmans and F. Mariz. 2023. The Blue Bond Market: A Catalyst for Ocean and Water Financing. *Journal of Risk and Financial Management*. 16(3). pp. 184.; N. Shiiba et al. 2022. How Blue Financing Can Sustain Ocean Conservation and Development: A Proposed Conceptual Framework for Blue Financing Mechanism. *Marine Policy*. 139.

4.4.2 Public–Private Partnership

Overview

Asia's infrastructure requirements are estimated at over \$1.7 trillion a year until at least 2030.[231] There is a huge gap between the financing that is required and what governments can afford to pay from their budgets. This shortfall—along with weak capacity, an underdeveloped enabling environment, poor project preparation, and insufficient project financing—is holding back economic growth and poverty alleviation. PPP thus offers a solution to reduce the gap between the financing that is required and what governments can afford to pay from their budgets.

PPPs present a framework that—while engaging the private sector—acknowledges and structures the role of government in ensuring that social obligations are met and successful sector reforms and public investments are achieved. PPPs are typically long-term contractual arrangements where a government partners with the private sector to build and run infrastructure, such as a road or bridge. PPPs can be used to finance and run traditional infrastructure, as well as railways, ports, water, sewage facilities, and renewable power systems.

A strong PPP allocates the tasks, obligations, and risks among the public and private partners in an optimal way. The public partners in a PPP are government entities including ministries, departments, municipalities, or state-owned enterprises. The private partners can be local or international and may include businesses or investors with technical or financial expertise relevant to the project. Increasingly, PPPs may also include NGOs and/or community-based organizations that represent stakeholders directly affected by the project.

Effective PPPs recognize that the public and the private sectors each have certain advantages, relative to the other, in performing specific tasks. The government and private sector may play the following roles in a PPP:

[231] ADB. *ADB and Public-Private Partnerships.*

(i)   The government's contribution to a PPP may take the form of capital for investment (available through tax revenue), a transfer of assets, or other commitments or in-kind contributions that support the partnership. The government also provides social responsibility, environmental awareness, local knowledge, and an ability to mobilize political support.

(ii)   The private sector's role in the partnership is to make use of its expertise in commerce, management, operations, and innovation to run the business efficiently. The private partner may also contribute investment capital depending on the form of the contract. The structure of the partnership should be designed to allocate risks to the partners who are best able to manage those risks and thus minimize costs while improving performance.[232]

Background

A few nations—like Chile, New Zealand, and the United Kingdom—privatized infrastructure utilities in the 1980s, and the preliminary results suggested that privatization would offer a remedy for underperforming public utilities. The argument was that a private company would function more effectively because of its profit-driven nature and having precise, consistent aims and means as opposed to the numerous and frequently incompatible purposes imposed on state-owned utilities. The 1989 privatization of water in England and Wales was a key factor in persuading decision-makers that financing urban water utilities privately would be an option. This privatization was a historic development for the sector globally, raising enormous amounts of private capital from global financial markets.

Access to private funds became the primary driver of governments seeking private sector engagement in urban water utilities as they faced significant investment demands. In 1993, significant momentum was created by the Buenos Aires concession grant. After acquiring control, the concessionaire had early success. During the first year of private operation, the city resolved its ongoing summer water rationing issues, and during the first 4 years of private operation, connected more than 1 million people to the water network, closing the gap with the average national coverage.[233]

The three main needs that motivate governments to enter into PPPs for infrastructure are

(i)   to attract private capital investment;

(ii)   to increase efficiency and use available resources more effectively; and

(iii)   to reform sectors through a reallocation of roles, incentives, and accountability.

PPPs personify a popular policy and financing option in many developing countries, and development banks like ADB and the World Bank have been instrumental in encouraging innovation and trying out market-oriented strategies in the water industry. With rising awareness of elements that can impede project design and implementation of PPPs, like crucial contract provisions and legal foundations (including regulation) to facilitate PPPs, there is an increased interest in using PPPs to address water-related problems in the region, including pollution.

Implementation

PPP is increasingly being implemented in diverse water subsectors in developing Asia. PPPs are commonly used in the water sector to develop, operate, and maintain various water infrastructure projects, including water distribution, wastewater treatment, and bulk water treatment and transmission. These partnerships involve collaboration

---

[232]   ADB. n.d. *Public–Private Partnership Handbook.*
[233]   P. Marin. 2009. *Public–Private Partnerships for Urban Water Utilities.* World Bank.

between public authorities and private sector entities to deliver efficient, sustainable, and cost-effective water services. The three major water subsectors where PPP contracts are becoming common are as follows:

(i) **Water Distribution**

    (a) **Concession contracts**: Under a concession contract, a private company is granted the right to operate and manage a water distribution system for a specified period. The private partner may be responsible for maintenance, infrastructure investment, billing, and customer service. A typical concession partnership can have a 20- to 40-year contract term, with the fee fixed for the term of the contract, which creates budget certainty.[234]

    (b) **Management contracts**: In a management contract, the private partner takes on the operational and management responsibilities of a water distribution system while the ownership remains with the public entity. This arrangement often focuses on improving system efficiency and reducing water losses.[235]

(ii) **Wastewater Treatment**

    (a) **Design-Build-Operate contracts**: In a Design-Build-Operate contract, a private entity is responsible for designing, building, and operating a wastewater treatment plant or system. These contracts often include performance-based criteria to ensure the treatment facility meets environmental standards.[236]

    (b) **Build-Operate-Transfer contracts**: In a Build-Operate-Transfer contract, a private company finances, builds, and operates a wastewater treatment facility to transfer it back to public ownership after a specified period.[237]

(iii) **Bulk Water Treatment and Transmission**

    (a) **Water supply contracts**: Private companies may enter into long-term contracts with public authorities to supply bulk water. These contracts can include the construction and operation of water treatment plants and transmission systems to deliver treated water to municipalities.[238]

    (b) **Offtake agreements**: In some cases, the private sector provides treated water to public utilities, industrial users, or agricultural clients through offtake agreements. These arrangements may specify the terms of water supply, quality standards, and pricing.[239]

According to data from the World Bank's Private Participation in Infrastructure (PPI) database, Asia executed water PPP project transactions totaling $29.5 billion during 2000–2019 and $12.6 billion during 2010–2019 (Table 6). The average size of a PPP water project transaction in Asia was $46.2 million during 2000–2019. Together, the PRC, the region of Latin America, and the Caribbean accounted for 82% of the total volume and 67% of the total value of PPP initiatives in the water and wastewater sector during 2000–2020. About 524 out of 638 PPP contracts in Asia during 2000–2019 were for water treatment plants. One hundred eight projects dealt with water distribution, and six projects included both water distribution and treatment.

---

[234] Investopedia. 2024. *Concession Agreement: What It Is and How It Works.*
[235] World Bank. 2021. *Management Contract—Water and Wastewater (Example 1)*
[236] ADB. 2021. *User guide for design–build–operate contracts for water and wastewater greenfield infrastructure projects.*
[237] Investopedia. 2023. *Build-Operate-Transfer Contract: Definition, Risks, and Framework.*
[238] World Bank. 2022. *Water and Sanitation PPPs.*
[239] World Bank. 2020. *Sample Bulk Supply Agreement (BOT) for Water.*

Table 6: Water Public–Private Partnerships Awarded Globally by Region, Volume, and Value

Regions	2000–2009				2010–2019			
	No. of projects	%	$ Million	%	No. of projects	%	$ Million	%
Asia	378	64	16,860	55	260	71	12,693	40
PRC	312	53	8,013	26	234	64	9,615	30
Rest of Asia	66	11	8,847	29	26	7	3,024	10
LAC	134	23	2,870	26	90	25	16,334	52
SSA	15	3	164	1	4	1	219	1
MENA	17	3	3,202	11	8	2	1,147	4
Europe	40	7	2,296	8	3	1	1,333	4
Total	**584**	**100**	**30,392**	**100**	**365**	**100**	**31,672**	**100**

LAC = Latin America and Caribbean, MENA = Middle East and North Africa, PRC = People's Republic of China, SSA = sub-Saharan Africa.
Source: World Bank PPI database.

Applicability in Southeast Asia

A PPP narrative that can be universally applied to all Southeast Asian countries does not exist, but one can discern certain recurring patterns in the context of what we might call water service systems. These systems involve the collaboration of various levels and branches of government along with private entities in the provision of water and wastewater services in the rivers, seas, and oceans. Consequently, within the relatively limited scope of PPP projects, it has been more challenging to implement PPPs with a focus on water distribution in Asia. This observation aligns with the broader institutional and financial limitations encountered in the Asian water sector (Table 7).

Table 7: Public–Private Partnerships in Asia by Volume and Value

Countries	Volume		Value	
	No. of projects	%	$ Million	%
PRC	546	86	17,629	60
Malaysia	11	2	6,502	22
India	20	3	1,258	4
Philippines	8	1	1,249	4
Indonesia	11	2	759	3
Thailand	16	3	544	2
Others	26	4	1,560	5

PRC = People's Republic of China.
Source: World Bank PPI database.

Southeast Asian nations like Malaysia and Thailand have implemented sector-level initiatives to promote PPPs. In Indonesia and the Philippines, considerable concessions are being granted. For instance, in the Philippines, the support for PPPs has seen periods of decline and resurgence. The country's initial venture into PPPs in the water sector was a 50-year lease (referred to as a franchise) for water supply in a portion of Angeles City in 1960. The Manila Concessions represent a hybrid form of PPP, where they blend a concession agreement with the presence of a regulatory authority responsible for periodically assessing and adjusting tariffs based on the principles and procedures outlined in the contract. Remarkably, the regulatory framework has proven to be resilient, even considering the challenges faced during the initial years of the concessions. As a result, these concessions have successfully expanded coverage and enhanced technical and commercial efficiency to a significant extent.[240]

Challenges and key success factors

Comparatively to Latin America—where PPPs have been supported by well-developed regulatory frameworks (e.g., Chile)—Southeast Asia's adoption of PPPs is hindered by a lack of institutional maturity and advancement of regulatory mechanisms. Local governments frequently regulate contracts and establish tariffs, even if they do not have the functional capability or financial resources to implement and oversee PPP projects.

There are three complementary pivots along which governments need to act to scale water PPPs and their development impact:[241]

(i)   The first is to put in place a holistic water governance framework that encompasses three actions:

(a)   public policy commitment to water security and inclusive access that is fiscally sustainable;

(b)   empowered and capable public counterparty institutions mandated with service delivery, contract monitoring, and regulatory functions; and

(c)   a progressive revenue regime where revenues are linked to usage and service, and subsidies are transparent, targeted, and formalized in advance.

(ii)   The second is to foster an enabling environment for PPPs comprising three aspects:

(a)   a sector-specific PPP strategy that identifies clear objectives and a pipeline of projects to aid programmatic implementation;

(b)   greater rigor in project preparation and closure; and

(c)   transparent fiscal support rules and consistent frameworks to manage and monitor fiscal costs and contingent liabilities.

(iii)   The third is to design transactions in the project that incorporate features for bankability and competitive tension and focus on outcomes through the following:

(a)   supply-side impetus, competition efficiency, and transparency of procurement;

(b)   a sharp focus on operations and maintenance, clear performance linkages, and post-award management;

(c)   contract sanctity and payment security; and

(d)   contextual fit and appropriateness.

Boxes 12 and 13 present examples of PPP projects.

---

[240]   O. Jensen. 2014. *Public–Private Partnerships for Water in Asia: A Review of Two Decades of Experience. International Journal of Water Resources Development. 33 (1).*

[241]   H. Rahemtulla et al. 2022. *A Governance Approach to Urban Water Public–Private Partnerships Case Studies and Lessons from Asia and the Pacific.* ADB.

Box 12: As-Samra Blended Finance Structure through Public–Private Partnerships

Background

The As-Samra Wastewater Stabilization Ponds—which were outdated and overburdened—were replaced by the new As-Samra Wastewater Treatment Plant (WWTP). The As-Samra WWTP was built during 2003–2008 for $169 million. Jordan's water quality was improved with the construction of the As-Samra WWTP. It managed the wastewater that is discharged from the Greater Amman and Zerqa metropolitan areas as well as the Zerqa river basin. Most of the polluted water from the Zerqa River is discharged into the King Talal Dam, which provides Jordan Valley irrigation water, posing serious environmental and health risks. Jordan's As-Samra WWTP expansion exemplifies how blended finance can attract private investment.

Implementation

The water treatment facility was expanded through a build-operate-transfer (BOT) agreement signed between the Ministry of Water and Irrigation and Samra Wastewater Treatment Plant Company Limited. The BOT contract included the design, building, finance, and 22-year operation of the facility:

(i)   The public–private partnership attracted private investors like Suez, Morganti, and Infilco Degrémont, in addition to the $93 million grant from the Millennium Challenge Corporation (MCC).

(ii) The Arab Bank obtained a grant of $20 million from the Government of Jordan and $110 million in private financing through a loan syndication process as a part of the varied blend of financing.

(iii) Swedish International Development Agency funding was supplied for the first 18 months of the commissioning phase's planning phase.

With this project, more water was made available for agriculture and other high-value uses by the communities.

Financial Structure for the As-Samra Wastewater Treatment Plant Expansion in Jordan

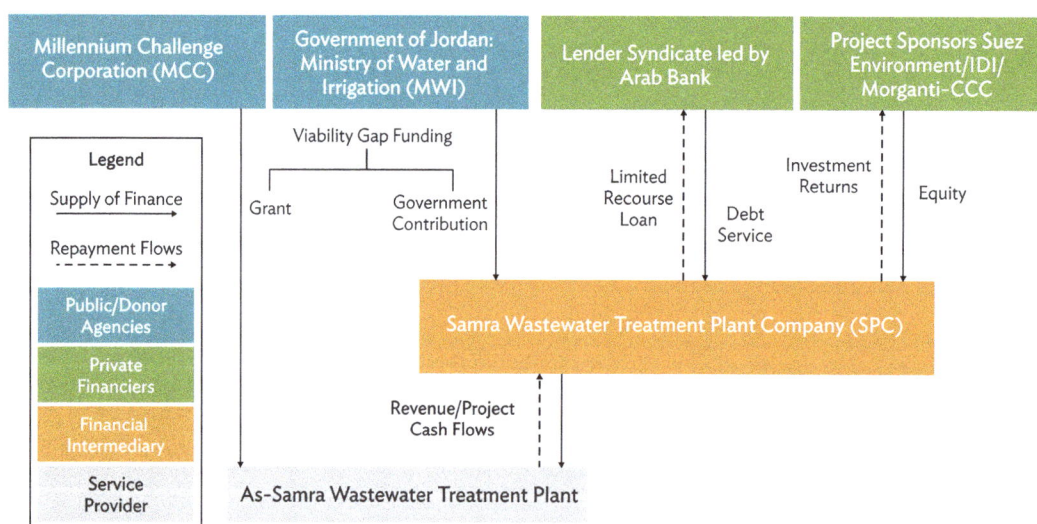

Source: World Bank Group. 2016. *Case Studies in Blended Finance for Water and Sanitation: Blended Financing for the Expansion of the As-Samra Wastewater Treatment Plant in Jordan.*

continued on next page

Box 12 *continued*

Key lessons learned

(i) **Funding the "viability gap" is crucial**. The MCC award reduced the project's capital expenses and made it financially feasible, which was crucial for the use of commercial financing.

(ii) **Donor requirements prompted nontraditional project finance agreements**. The only types of project financing that are typically available are debt and equity; however, in this case, the addition of grant financing made certain donor funding conditions for the project necessary, leading to more complicated negotiations than in a typical project finance transaction. To maintain environmental sustainability, it was crucial to abide by the tight regulations set by Jordan and MCC for the management, storage, and disposal of sludge. Ultimately, the creative approach to combining funding sources made the project a success, where all parties expressed their satisfaction at the financial close.

Sources: Water Technology. *As-Samra Wastewater Treatment Plant, Jordan*; OECD. 2018. *Background Paper: Blended Finance for Water-Related Investments*; World Bank Group. 2016. *Case Studies in Blended Finance for Water and Sanitation: Blended Financing for the Expansion of the As-Samra Wastewater Treatment Plant in Jordan*.

Box 13: Moya and PAM JAYA Water Cooperation Agreements in Jakarta, Indonesia

Background

Land subsidence is a major coastal threat in Jakarta, Indonesia, and the city is sinking by an average of 12–18 centimeters every year. The Minister of Public Works and Housing—Basuki Hadimuljono—revealed that the main cause of the problem is the overuse of groundwater. Many commercial buildings, industrial entities, and residences built their underground wells as their main water source. The number of underground wells in housing areas resulted in a massive space underground. This then creates a major subsidence problem, and the government is unable to ban groundwater usage before providing a solution.

Moya Holdings Asia Limited announced that the company's wholly owned subsidiary—PT Moya Indonesia—had, in October 2022, been awarded projects that were put up for tender by PAM JAYA, a municipal water company in Jakarta. Under the tender, PAM JAYA intended to enter into three cooperation arrangements with a private water treatment company to optimize the existing water assets ("Project Brownfield") and to build new water assets ("Project Greenfield") in Jakarta on behalf of PAM JAYA.

Implementation

PT Air Bersih Jakarta (a subsidiary of PT Moya Indonesia, referred to as "ABJ") entered into a credit agreement with a syndication of PT Bank Central Asia Tbk, PT Bank OCBC NISP Tbk, PT Bank Tabungan Negara (Persero) Tbk, PT Bank BTPN Tbk, PT Sarana Muli Infrastruktur (Persero), PT Bank KB Bukopin Tbk, and PT Bank China Construction Bank Indonesia Tbk for banking facilities of up to a total principal amount of Rp8.8 trillion. The projects would have a 25-year cooperation period, which was expected to start when operational management of the existing water treatment plants commences in May 2023.

continued on next page

Box 13 *continued*

(i) The scope of Project Brownfield is the rehabilitation, operation, and maintenance of the existing water treatment plants under the rehab-operate-transfer scheme.

(ii) The syndication agreement was arranged by PT Bank Central Asia Tbk, Oversea-Chinese Banking Corporation Limited, and PT Bank OCBC NISP Tbk. The signing of this credit agreement with the syndication lenders signified a key milestone under the three agreements.

(iii) PT Moya Indonesia targets to install about 2,500 kilometers of pipe network by 2025 in Jakarta. It also aims to install about 350,000 new pipe connections.

(iv) ABJ will build and operate the clean water infrastructure, while the downstream activities will be carried out by PAM JAYA.

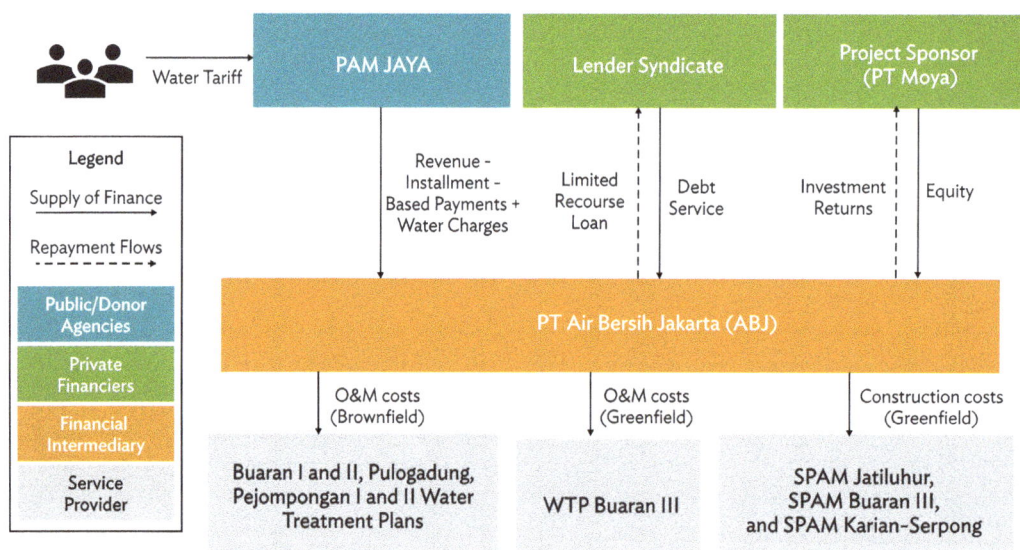

Delivery of Water Treatment and Distribution Assets in Jakarta, Indonesia

O&M = operation and maintenance, SPAM = Regional Drinking Water Provision System, WTP = water treatment plants
Sources: Moya Asia. 2023. Moya Group secures IDR 8.8 trillion credit facility to accelerate 350,000 new piped water connections in Jakarta. *Press release*. 21 February; EY Analysis.

Impact

(i) **Minimize groundwater usage**. The new credit facility is expected to accelerate the projects with which Moya Group believes that piped water access can become the solution to stop groundwater usage that can prevent the further sinking of Jakarta.

(ii) **Improve water service coverage**. The combined projects aim to help PAM JAYA reach 100% service coverage by 2030.

continued on next page

Box 13 *continued*

Key Takeaways

  (i) **Public–private partnerships are financed and procured through nonsovereign entities**. While there is often a national, subnational government, or municipality involved in a PPP transaction, there are opportunities for nonsovereign entities to play a key role in PPPs. In this case, the project tender was called by PAM JAYA, the regional government-owned water supply company responsible for providing water in Jakarta.

  (ii) **Significant appetite for water projects from private sector investors, both strategic and financial**. Water infrastructure remains attractive to private sector investors because of the essential nature of water services and the corresponding long-term revenue generation potential.

  (iii) **Incorporation of both greenfield and brownfield elements**. Greenfield elements in water PPP projects involve the development of entirely new water infrastructure or facilities, while brownfield elements involve the redevelopment, rehabilitation, or expansion of existing water infrastructure. These projects are often aimed at modernizing, upgrading, or optimizing the performance of aging water assets. The water cooperation agreements between PT Moya Indonesia and PAM JAYA involve both optimizing existing water treatment plants and building a new water distribution network.

IDR = Indonesian rupiah, PPP = public–private partnership, Rp = Indonesian rupiah
Sources: Mckinsey. 2014. Using PPPs to fund critical greenfield infrastructure projects; Moya Asia. 2023. Moya Group secures IDR8.8 trillion credit facility to accelerate 350,000 new piped water connections in Jakarta. *Press release*. 21 February; OECD. 2018. Background Paper: Blended finance for water-related investments.

4.4.3 Blended Finance

Overview

Blended finance refers to the use of catalytic capital from public and philanthropic sources to increase private sector investment in sustainable development. Blended strategies seek to raise more money for investments in sustainable development in developing nations and can serve as a method to lower risk and boost lender confidence. Blended finance functions as a market-building instrument that creates a transition from dependence on grants and other donor finance toward commercial financing. It does so by deploying development finance in a way that addresses investment barriers that prevent commercial investment in SDG-relevant sectors, such as water and sanitation.

Blended finance transactions have three signature markings:

  (i) **Development impact and Sustainable Development Goals**: The financing should contribute to attaining the SDGs.

  (ii) **Return**: Positive financial return can be expected.

  (iii) **Leverage**: Philanthropic parties are catalytic, leading to an increased risk-reward profile that can mobilize and/or attract "additional" private sector investment.[242]

Blended finance can be sourced from a wide range of investors such as from government, philanthropic funds, international financial institutions, NGOs, multilateral agencies, and scientific institutions. Even private finances such as equity investors, impact investors, commercial banks, pension funds, and crowdfunding can be potential

---

[242] K. Sachwani. 2020. *Financing the Water Sector—An Alternate Approach*. World Bank Group.

sources for financing. Blended finance is a structuring approach that allows organizations with different objectives to invest alongside each other while achieving their objectives (whether financial return, social impact, or a blend of both). The main investment barriers for private investors addressed by blended finance are (i) high perceived and real risk and (ii) poor returns for the risk relative to comparable investments. Blended finance creates investable opportunities in developing countries that lead to more development impact.[243]

Background

According to Convergence—a global membership network for blended finance—blended finance has mobilized about $198 billion in capital toward sustainable development in developing countries (Figure 26). Blended finance transactions range considerably in size, from a minimum of $110,000 to a maximum of $8 billion. Based on transactions that occurred during 2010–2018, the median blended finance transaction has been $64 million (footnote 243).

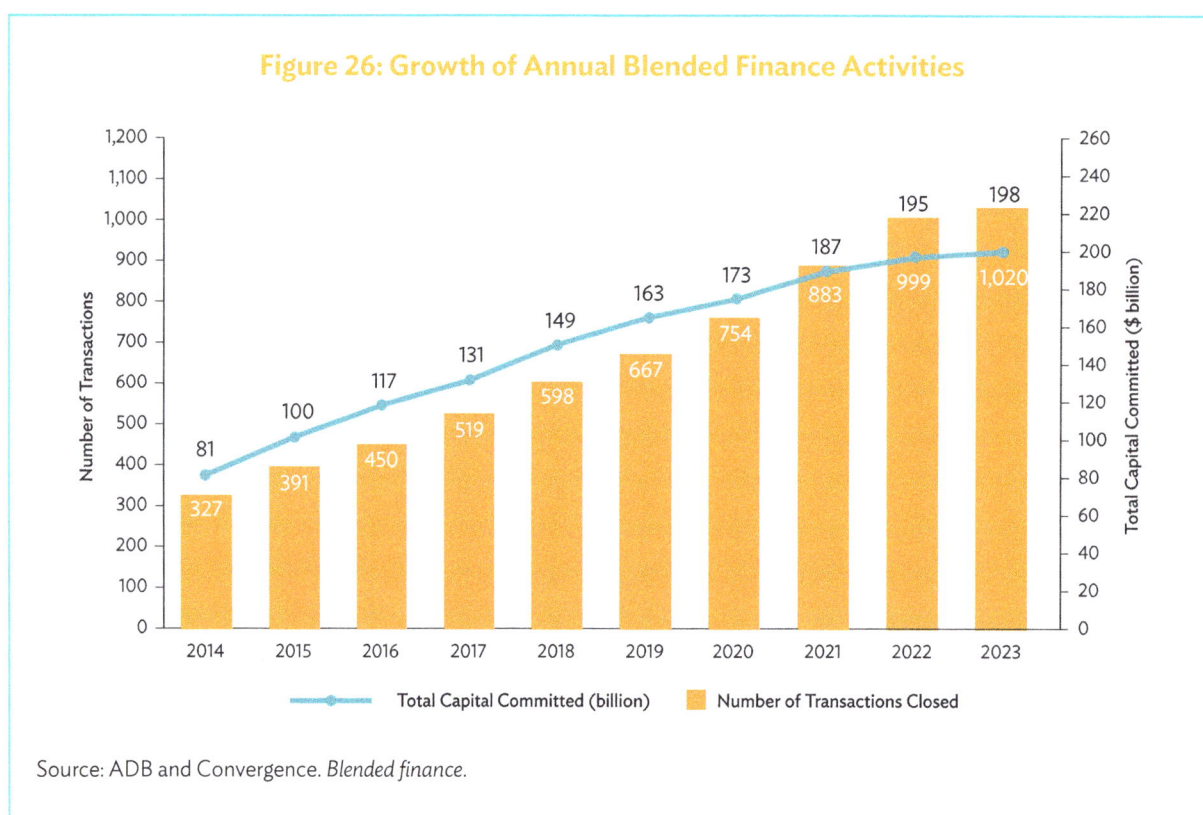

Figure 26: Growth of Annual Blended Finance Activities

Source: ADB and Convergence. *Blended finance.*

Sub-Saharan Africa has been the most frequently targeted region in blended finance transactions, although Asia and Latin America have been emerging as new frontiers for blended finance. More than 1,800 unique investors have participated in one or more blended finance transactions, and nearly two-thirds of these investors come from the private sector. SDG 17 (Partnerships for the Goals), SDG 8 (Decent Work and Economic Growth), SDG 09 (Industry, Innovation, and Infrastructure), and SDG 1 (No Poverty) have been the most frequently targeted in blended finance transactions (footnote 243).

---

243 Convergence. Blended finance.

Despite growing interest from both public and private actors, commercial finance mobilized for the water sector accounts for less than 1.5% by value and only 5% of total transactions by volume. This implies that the water sector has so far attracted only a small share of blended finance, although the requirement for investment is substantial.

For instance, an analysis of the expenses associated with reaching the SDG Target 6.2 (Clean Water and Sanitation) goal alone estimates that the yearly costs will be close to $70 billion during 2017–2030. Including the capital requirements for achieving SDG 6.1 (Safe and Affordable Drinking Water), the investment requirements are as high as three times the existing levels of commitment. This deficit cannot be filled with public and philanthropic finance alone.[244]

Implementation

Several SDGs—particularly those about food security, healthy living, clean energy, and marine and terrestrial ecosystems—are impacted by investments in the water sector. This reflects the wide range of water-related expenditures and the numerous demands that these investments can meet (such as ensuring a steady supply of fresh water, lowering pollution, supplying drinking water, offering sanitation and wastewater treatment services, and irrigation).[245] Blended financial strategies can be categorized based on the instruments and methods used in Figure 27.

Figure 27: Instruments to Mobilize Private Capital as Part of Blended Financing

Mechanism
Structure and/or intermediation of instruments to mobilize private capital

Funds

Syndication

Securitization

PPPs

Instruments
Mitigating risk and crowding in additional capital

| Equity Instruments | Debt Instruments | Mezzanine Instruments |
| Guarantees and Insurance | Hedging | Grants and Technical Assistance |

PPP = public–private partnership.
Source: OECD. 2019. *Making Blended Finance Work for Water and Sanitation: Unlocking Commercial Finance for SDG 6.*

Mobilizing private capital is a critical step in blended financing. Several instruments and strategies such as PPP, concessional finance, impact bonds, and debt instruments can be employed to attract and leverage private capital as a part of blended financing. To effectively mobilize private capital, it is crucial to tailor the approach to specific project needs, with consideration for risk management and alignment of incentives for all stakeholders involved.

---

[244] World Water Council. 2022. *Blended finance in the water sector.*
[245] OECD. 2019. *Making Blended Finance Work for Water and Sanitation: Unlocking Commercial Finance for SDG 6.*

Public sector organizations, development institutions, and private investors should work together to create sustainable and impactful financing structures.

Key stakeholders involved in the process of blended finance include investors and various types of intermediaries (Figure 28):[246]

(i)   **Investors**: Two main types of investors provide long-term, large-scale capital to fund critical development needs:

   (a)   **Institutional investors** (such as banks, insurers, or asset managers) invest the vast majority of the required capital and typically seek profitable risk-adjusted returns.

   (b)   **Concessionary investors** (public development assistance and foundations) make smaller investments but may be willing to accept a higher risk of loss or receive returns that are lower than the market rate.

(ii)   **Intermediaries**: Teams of financial experts who match projects with investment capital are frequently funded by international development banks or the private sector (foundations and NGOs).

(iii)   **Projects**: Funding secured is used for specific needs often found in emerging markets including clean water access, medical facilities, water treatment, and sustainable development programs.

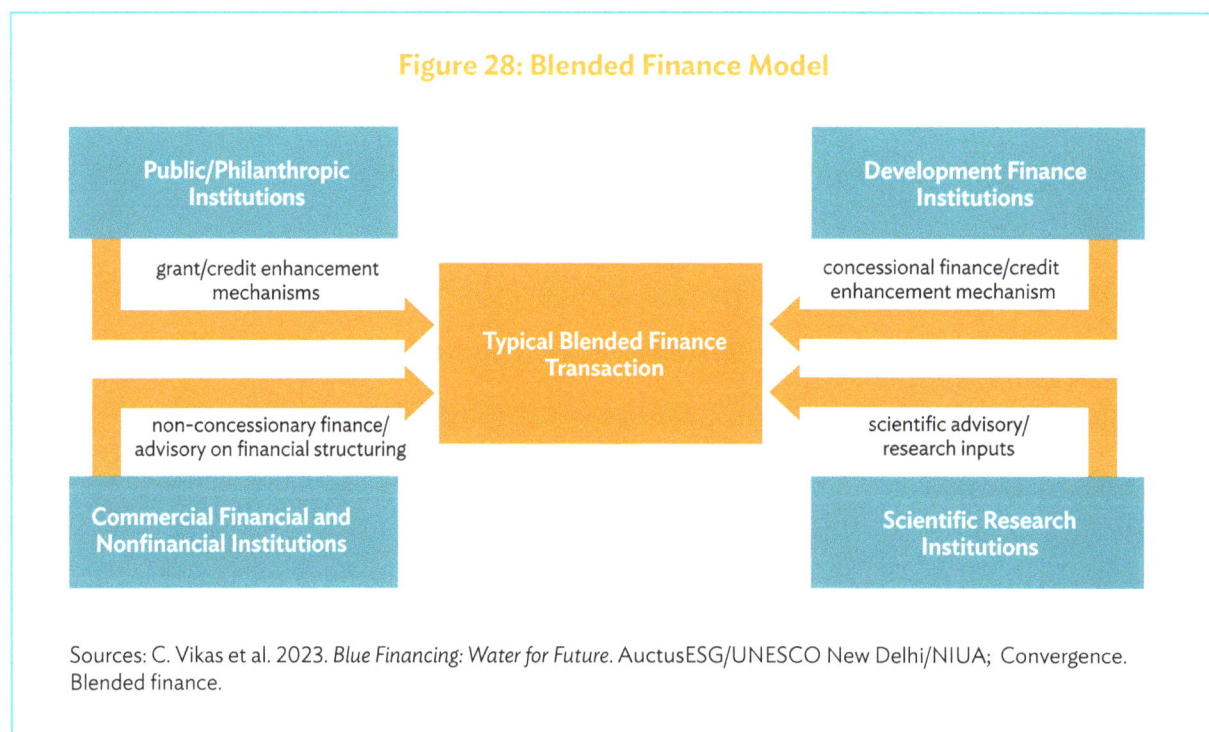

Figure 28: Blended Finance Model

Sources: C. Vikas et al. 2023. *Blue Financing: Water for Future.* AuctusESG/UNESCO New Delhi/NIUA; Convergence. Blended finance.

There are several benefits to using blended finance to address infrastructure financing gaps:

(i)   **Effectively address market failures and improve the commercial feasibility of projects.** Blended finance involves the use of public or philanthropic capital to provide risk mitigation instruments such as guarantees or insurance. By addressing market failures and mitigating risks, blended finance can attract private sector investment to sectors and regions that might not otherwise receive sufficient funding (footnote 245).

---

[246]   Bank of America. 2023. *What is blended finance, and why it matters.*

(ii) **Risk tolerance flexibility**. Blended finance enables investors to select varying risk tolerances while collaborating on the same project. Investors who need to take less risk but are interested in funding high-impact projects can benefit from the capital cushion provided by those who are willing to accept more risk. This degree of flexibility is especially important in developing countries, where sustainable finance may have a significant impact on social and environmental issues (footnote 246).

(iii) **Credit enhancements by providing finance at lower interest rates and by minimizing default risk**. Blended finance can encourage private sector participation by reducing the risk of default and increasing investor confidence. By leveraging public funds to provide guarantees or insurance, blended finance can enhance the creditworthiness of projects, thereby allowing access to financing at lower interest rates, reducing the overall cost of capital, and improving financial viability.[247]

(iv) **Technical assistance, capacity building, and knowledge transfer**. Blended finance helps project developers and implementing agencies build the necessary skills and expertise to plan, implement, and manage projects effectively. Enhanced technical expertise leads to better project design, more accurate risk assessment, and more efficient implementation, reducing the likelihood of costly delays or errors.[248]

(v) **Early stage development and design funding of projects**. Blended finance can allocate funds to design projects including feasibility studies, environmental and social impact assessments, engineering and architectural design, and other preparatory work. Design funding can support innovative and pioneering projects, fostering the creation of new models and approaches to development (footnote 246).

Challenges

Three key challenges surface in the application of blended finance:

(i) **Difficulty in understanding the needs and incentives of multiple stakeholders**. Blended finance often involves a range of stakeholders, including governments, development finance institutions, private investors, project developers, and local communities. Each of these stakeholders may have different objectives, timelines, and risk appetites.

(ii) **The evaluation framework must be tailored to different types of investment instruments.** Blended finance encompasses various financial instruments including grants, concessional loans, guarantees, equity investments, and technical assistance. Each instrument has different risk profiles, return expectations, and impact measurement requirements. The evaluation framework needs to account for risk assessment and mitigation strategies specific to each instrument. Evaluating the success of blended finance projects in achieving their intended social, economic, and environmental impact is a complex task.[249]

(iii) **The need to accurately measure and demonstrate the true additionality of blended finance**. Additionality refers to the concept that blended finance should provide financial resources and support that are additional to what would have been available from purely public or purely private sources. Demonstrating additionality is crucial for ensuring that blended finance effectively mobilizes additional resources and achieves its intended impact (footnote 245).

Examples of blended finance instruments are presented in Boxes 14 and 15.

---

[247] OECD. 2022. *Scaling Up Blended Finance in Developing Countries.*
[248] World Bank Independent Evaluation Group. 2020. What is blended finance, and how can it help deliver successful high-impact, high-risk projects?
[249] Observer Research Foundation (ORF). 2023. *Enhancing Blended Financing for a Sustainable Future: Challenges and Potential Solutions.*

Box 14: Republic of Seychelles Blue Bond Issuance Using Debt Swap

Background

As it is a small, developing island nation, the economy of Seychelles is inextricably linked to its marine and coastal resources. The ability of the nation to commit finances for climate adaptation initiatives to protect and preserve its marine resources was severely hampered by the 2008 financial crisis. In 2018, the Republic of Seychelles introduced the first sovereign Blue Bond and undertook a debt-for-nature swap aimed at specifically protecting the world's oceans and biodiversity.

The Blue Investment Fund (run by the Development Bank of Seychelles) and the Blue Grants Fund (run by the Seychelles Conservation and Climate Adaptation Trust [SeyCCAT]) gave loans and grants. The bond—which raised $15 million from foreign investors—showed the potential of using capital markets to finance the wise exploitation of marine resources. The bond issue was coupled with an additional $10 million of funding, consisting of a $5 million grant from the Global Environment Facility and $5 million in low-interest loans from the International Bank for Reconstruction and Development (IBRD). The bond used debt swap to support issuance by undertaking the world's first Blue Economy debt for nature swap, to convert $21.6 million of the national debt.

Impact

The bond was not issued for ocean conservation but aimed at achieving broader blue economy objectives.

(i) It was used to expand sustainable-use marine protected areas inside the exclusive economic zone of Seychelles that are subject to fishing regulations.
(ii) It also aimed to improve the governance of priority fisheries and develop the institutional capacity to improve the governance of priority fisheries.
(iii) It focused on sustainably developing the blue economy by increasing investment in aquaculture, industrial, semi-industrial, and artisanal fishing.
(iv) SeyCCAT was to oversee perpetual endowment and enforce the provisions of the debt restructuring agreement in addition to providing funding for climate adaptation and marine conservation initiatives. It was also in charge of giving the government and NGOs their share of the debt swap earnings.

Benefit to the government

(i) Redirection of external debt service to investments in the country.
(ii) Improved fiscal space because of extended maturities on $21.6 million of debt from 8 years to 13 years on average.
(iii) Government entities were eligible to apply for funding from SeyCCAT.

Implementation

The key stakeholders involved were the Government of Seychelles, The Nature Conservancy, the Paris Club of Creditors, the World Bank Group, and the Global Environment Facility (GEF):

(i) IBRD provided a loan guarantee for the $15 million Seychelles Blue Bond.
(ii) The bond offering was accompanied by an extra $10 million in funding, which was made up of $5 million in low-interest loans from IBRD and $5 million in grants from the GEF (which will help pay some of the bond interest costs).

continued on next page

Box 14 *continued*

(iii) SeyCCAT raised $20.2 million through a hybrid financing model. Among them, $5 million came from charitable donations, and another $15.2 million came from the securities investment of influential investors.

(iv) Seychelles paid bondholders from the central budget.

Framework of Seychelles Innovative Financing for the Blue Economy

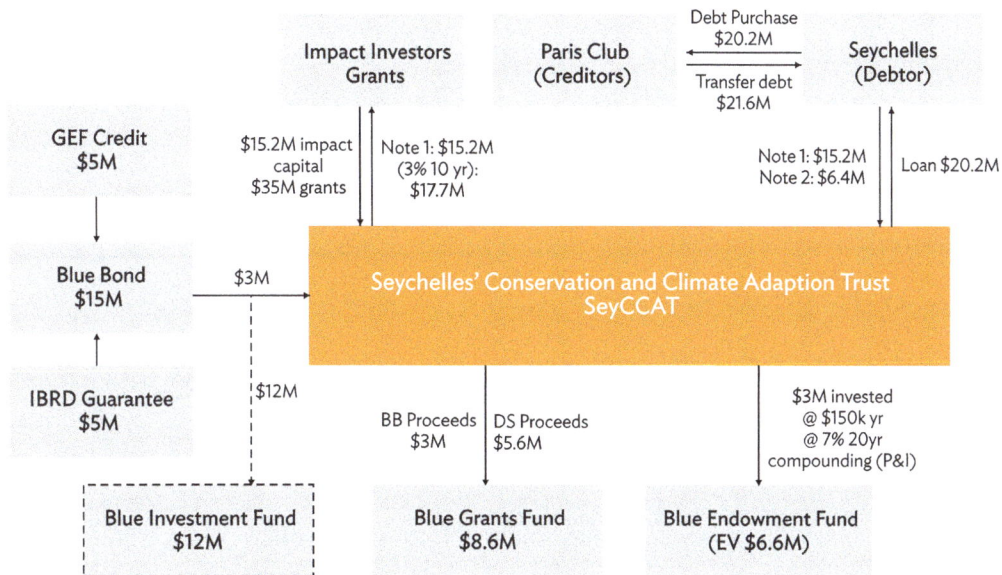

BB = blue bonds, DS = debt swap, EV = expected value, GEF = Global Environment Facility, IBRD = International Bank for Reconstruction and Development, M = million, SeyCCAT = Seychelles Conservation and Climate Adaptation Trust, yr = year.

Source: The Commonwealth. 2020. *Case Study: Innovative Financing—Debt for Conservation Swap, Seychelles' Conservation and Climate Adaptation Trust, and the Blue Bonds Plan, Seychelles* (ongoing).

Key Takeaways

(i) **Seek sound and worthwhile investments**. Finance initiatives must be sufficiently significant to be worthwhile, but not so large as to be too great a risk. Risk can only be minimally absorbed by small economies like those in Seychelles and many other island countries. Greater investments entail a greater risk of loss. A thorough risk assessment is required before launching financing activities.

(ii) **Invest in a robust administrator and ensure monitoring**. For the program to continue to be innovative and for the money to reach the intended audience, there needs to be stable, long-term funding to support a skilled administrator. Monitoring is crucial to ensure project funds are used effectively.

continued on next page

Box 14 *continued*

(iii) **Implement social and environmental safeguards**. As a condition of receiving funding support from the World Bank and its partners, certain criteria must be met. The application of funds needs to be well thought out to avoid any unforeseen repercussions. The role of the Blue Grants Fund in fostering new businesses is crucial, but in some circumstances, this may have the potential to increase environmental pressure, which runs counter to the concept of a sustainable blue economy, so caution and oversight must be exercised.

Sources: Y. Wang and Z. Wang. *Blue Finance Case Study—The Republic of Seychelles' innovative use of Debt for Nature Swap promotes marine protection.* International Institute of Green Finance (IIGF); V. Iyer et al. 2018. Finance Tools for Coral Reef Conservation: A Guide. Wildlife Conservation Society; The World Bank. 2019. *Thematic Bond Advisory: Blue Bond*—CASE STUDY; A. Pouponneau. 2023. *Applicability of innovative finance for addressing loss and damage.* United Nations Framework Convention on Climate Change (UNFCCC); The Commonwealth. 2020. *Case Study: Innovative Financing—Debt for Conservation Swap, Seychelles' Conservation and Climate Adaptation Trust and the Blue Bonds Plan, Seychelles* (ongoing).

Box 15: Water Revolving Fund in the Philippines

Background

In the past, international development funding, domestic public funds, and bill collection proceeds have all been used to finance the water and sanitation sector in the Philippines. Such revenues were found to be insufficient to pay for infrastructure investment expenditures in the 1990s, which raised interest in securing private-sector funding. Private bank funding to water utilities was, however, essentially nonexistent at the time since local commercial banks did not understand water utilities and viewed them as weak and ineffective.

The Philippine Water Revolving Fund (PWRF) was established in 2008 to offer loans to local government units and water districts that provide water services to support regional water and wastewater projects. Repayments on loans granted to the fund are rotated to fund other initiatives.

The PWRF combines commercial financing, domestic public money, and official development assistance (ODA) to reduce borrowing costs and promote sanitation and water projects to private finance institutions. The PWRF received concessional seed funding from the Japan Bank for International Cooperation (JBIC) at the program's beginning. A standby credit line (to guard against financial risk) and a co-guarantor credit guarantee are used to support the financing.

Impact

PWRF initially made a loan to the Puerto Princesa City Water District in 2010, where the money was used to repair and expand the water system of the city. The third-largest commercial bank in the Philippines—the Bank of the Philippine Islands—and the government-owned Development Bank of the Philippines jointly funded the loan, and

continued on next page

Box 15 *continued*

the cofinancing arrangement called for a loan of $13 million. The PWRF successfully channeled more than $234 million in financing for water and sanitation projects from its founding through 2014, with over 60% of these loans coming from private banks.

Implementation

To facilitate private institutional finance to support municipal water and wastewater projects, the US Agency for International Development (USAID) and JBIC worked in partnership with the Government of the Philippines to create the PWRF and help mobilize additional domestic commercial financing for water utilities through blending with ODA funding.

Philippine Water Revolving Fund Financial Structure

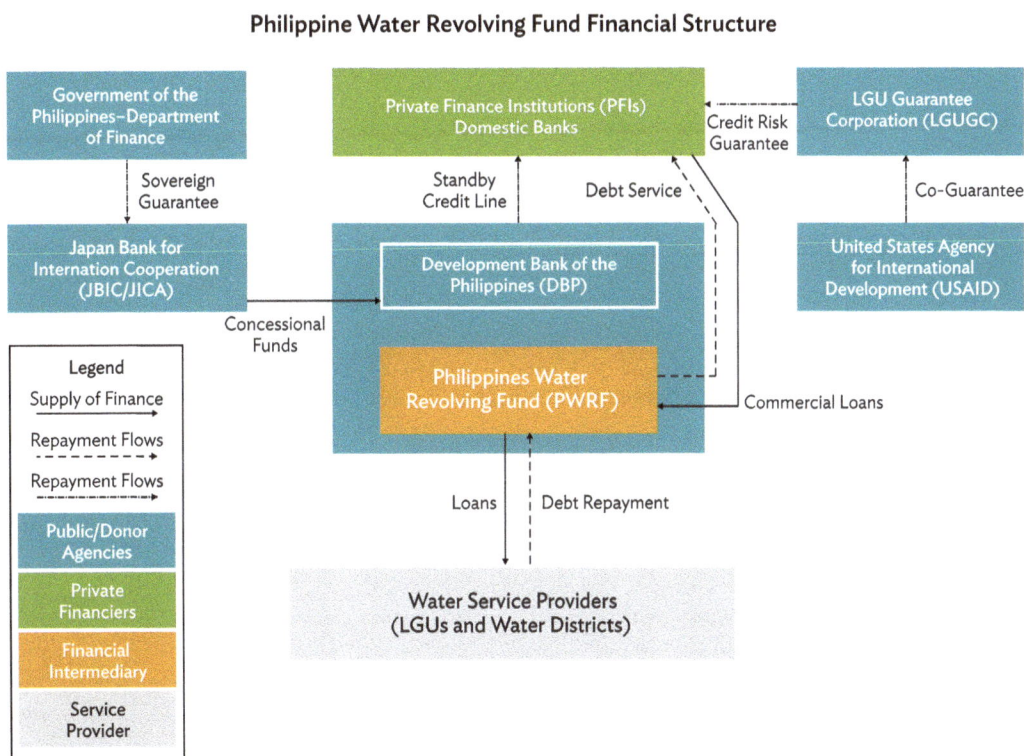

JICA = Japan International Cooperation Agency, LGU = local government unit.
Source: World Bank Group. 2016. Case Studies in Blended Finance for Water and Sanitation: Water Revolving Fund in the Philippines.

Key Takeaways

(i) **Longer tenors and cheaper borrowing costs for water service providers result from blending through the revolving fund**. By using standby credit lines, longer tenors can be achieved. PWRF incorporated a liquidity risk protection mechanism that gave private finance institutions the option to pay off their loan early by securing refinancing through the standby credit line from the Development Bank of the Philippines to extend loans that reflected the lifetime of utility assets. As a result, the utilities were able to obtain finance without having to significantly raise rates.

continued on next page

Box 15 *continued*

(ii) **Investment risk was reduced by the various credit enhancements offered with PWRF**. The PWRF blending design—made possible by the various credit enhancements—provides a vehicle for local private financial institutions to invest in the sector with little risk, revolutionizing how banks evaluate and fund water projects. However, the availability of such tools varies by nation (local government unit guarantee corporation-type entities do not exist in other nations) and from donors like USAID or the Development Credit Authority.

(iii) **Multilayered approach of PWRF to mobilize commercial finance is key**. Blended finance was accomplished through the revolving fund mechanism, but additional technical assistance and regulatory changes improved the business environment and access to commercial capital. Investors and utility companies received technical assistance to help them evaluate risk and improve performance.

Source: World Bank Group. 2016. Case Studies in Blended Finance for Water and Sanitation: Water Revolving Fund in the Philippines.

4.4.4 Other Potential Financing Mechanisms

Other financing mechanisms are widely used to minimize water pollution and to establish wastewater treatment infrastructure. These include subsidies, microfinance, partnerships, conservation trust funds, output-based aid, and sustainability-linked loans supporting water initiatives aimed at prevention, mitigation, and remediation. Each of these financing options has its strengths and can be adapted to the specific needs and context of river pollution control projects. A combination of these mechanisms—depending on the nature and scale of the pollution problem—can contribute to a comprehensive approach to tackling river pollution.

(i) **Subsidies**: Subsidies and grants are important features in most of the financing mechanisms since they provide viability gap funding. Government subsidies can be provided to industries, communities, and individuals to incentivize pollution reduction measures. For example, subsidies can support the adoption of cleaner technologies, the upgrade of wastewater treatment facilities, or the implementation of best practices in agriculture to reduce runoff. In Kitakyushu, Japan, the central government provided subsidies to the city government for the construction of sewer lines and wastewater treatment plants.

(ii) **Microfinance**: Microfinance institutions can provide small loans to local communities and small businesses for pollution control initiatives. This can include financing projects for proper waste disposal, sanitation, and eco-friendly agricultural practices. The targets of microfinance are usually low-income households that have no access to sanitation facilities. Examples of microfinance are in Cambodia, where microfinance loans for sanitation are handled by microfinance institutions, and the sanitation revolving fund in Viet Nam that provides loans to low-income households and is managed by the Women's Union.

(iii) **Output-based aid**: Output-based approach is a results-based financing mechanism where payments are made to service providers based on the delivery of predefined outputs or outcomes. It can be applied to projects that aim to reduce specific pollutants in rivers, and payments can be linked to the achievement of pollution reduction targets.[250]

---

[250] ADB. 2016. *Financing Mechanisms for Wastewater and Sanitation.*

(iv) **Sustainability-linked loans**: Sustainability-linked loans are financial instruments where the terms of the loan are tied to the achievement of predefined sustainability targets. Businesses and organizations can use these loans to fund pollution reduction projects and are incentivized to meet their environmental goals to secure favorable loan terms.[251]

(v) **Conservation trust funds**: Conservation trust funds are private, legally independent institutions that provide sustainable financing for sustainable activities. Conservation trust funds can be established to provide a sustainable source of financing for river pollution control efforts. These funds can be capitalized through public and private contributions and income generated from investments. They can support long-term environmental initiatives.[252]

---

[251] ICMA. 2019. *Sustainability linked loans principles.*
[252] Conservation Finance Alliance. *Practice standards for conservation trust funds.*

5 RECOMMENDATIONS FOR SOUTHEAST ASIA AND THE GREATER MEKONG SUBREGION

E Smart Bangkok Mass Rapid Transit Electric Ferries Project. The Project involves development and operation of about 27 fully electric ferries for mass public transport along the Chao Phraya River in Bangkok, Thailand. The service will be operated by E Smart and will operate on a 30-kilometer stretch of the Chao Phraya river in Bangkok and Nonthaburi province. (photo by Somkiat Jaraspat/ADB).

Water quality remains one of the most pressing environmental concerns faced globally and in Southeast Asia. The region has immense potential that can be unlocked through the prevention and reduction of river pollution. These benefits have far-reaching implications for the economy, people's livelihoods, health, biodiversity, and climate change. Based on the assessments made in previous chapters of this report, Southeast Asia has the potential to harness various nature-based solutions, policy measures, and institutional arrangements that have been successfully adopted in other regions.

However, to effectively implement these solutions, a broad range of challenges must be overcome, including strengthening technical expertise, monitoring and reporting capabilities, policy support, institutional setups, public awareness, and public participation. Improving financing is a catalyst that enables the implementation of all the solutions identified in this report. Yet, implementing and accessing these financial options also come with challenges such as the following:

(i) Identifying revenue streams for river pollution projects can be challenging.

(ii) Revenue streams for river pollution projects may not be sufficiently large.

(iii) River pollution projects often need to compete with other projects for capital.

This closing chapter shares recommendations for the next steps to be taken by key stakeholders in the region to develop projects that address river pollution challenges, along with a discussion about pilot projects that can ultimately lead to the development of robust and scalable solutions.

5.1 Pursuing Steps to Catalyze River Health Projects

River health relates to the condition and viability of ecosystems in river corridors (or land adjacent to rivers), and maintenance of river health depends on river flows, water quality, and the general environment of these corridors. Countries in the region can consider undertaking five key steps to catalyze projects to combat river pollution:

(i) **Consider a range of solutions**. There is no one-size-fits-all solution and municipalities need to consider utilizing a full range of options because each municipality faces unique challenges that may necessitate the implementation of more than just one locality-based solution. For instance, a river dealing with high nutrient pollution from agricultural runoff would benefit greatly from solutions such as wetlands or riparian buffers that can absorb the nutrients before they reach the river. This differs from a locale dealing with industrial waste from textile manufacturing, where using water pollution taxes as a form of mitigation might create a more direct impact. Therefore, municipalities should not dismiss any potential solutions prematurely.

(ii) **Develop comprehensive proposals to attract funding**. There is often a lack of well-developed project proposals that can appeal sufficiently to funders. Such proposals require a clear definition of the problem statement, the scope of work to be undertaken (e.g., technical assistance, capacity building activities, impact monitoring, and evaluation), and strong budget justification. To support the development of comprehensive proposals, there is a need to scale up the capabilities of local officials through training provided by multilateral organizations, river commissions, NGOs, and other development aid institutions. In particular, multilateral organizations such as ADB can directly support the development of these proposals through rigorous consultations with local stakeholders to properly identify responsible parties, clarify accountability, and ensure early stage buy-in of the proposal.

(iii) **Explore the use of new sources of capital**. Governments should explore the possibility of unlocking new sources of capital and combining them to fund pilot projects. This approach can stimulate research and development for a variety of pollution control strategies. For instance, a government could collaborate with environmental organizations, secure grants from international bodies, or issue blue bonds to raise funds for pilot projects aimed at testing and refining pollution mitigation methods. Pooling diverse sources of capital is crucial in obtaining the necessary funding for pilot projects, which can subsequently evolve into full-scale implementation.

(iv) **Solicit feedback from financiers**. Governments should solicit feedback from financiers to understand their requirements and incorporate their insights when it comes to scaling up pilot projects. It is crucial for governments to actively engage with financiers and seek their input on project priorities and requirements during the expansion of pilot initiatives. For example, a government running a pilot project aimed at compensating farmers for sustainable agricultural practices to reduce runoff pollution could involve private investors in discussions. This collaboration can lead to the development of scalable, financially attractive solutions that not only effectively address river pollution but also align with financier interests in terms of return on investment and environmental impact.

(v) **Foster cooperation at all levels of government**. Fostering cooperation at all levels of government—from local to international—is imperative because of the multifaceted and interconnected nature of the problem. River pollution often originates from various sources that transcend administrative boundaries, making it vital for governments to collaborate and share resources, knowledge, and strategies. For instance, a national government may collaborate with local authorities to develop and enforce stricter industrial discharge regulations along a polluted river. Simultaneously, international cooperation can be vital in addressing cross-border pollution issues, as the contamination of a river in one country can have far-reaching environmental and economic consequences for neighboring nations. This multilevel governmental collaboration is critical for effectively combating river pollution, as it allows for comprehensive solutions that consider the complexity and interconnectedness of the issue.

5.2 Developing Pilot Projects to Assess the Feasibility of Solutions

The importance of water security to broader economic development and resilience is increasingly recognized. Investments that contribute to water security span a range of essential infrastructure systems to deliver clean drinking water and reliable sanitation, as well as to manage water resources and water risks. Closing the financing gap for water-related investments—which remains key to addressing the problem of river pollution in Southeast Asia—requires mobilizing additional sources of finance and funding from various public and private sources. In many countries, the lack of a bankable, investment-ready pipeline of infrastructure projects is one of the major bottlenecks in attracting private capital to infrastructure. To successfully attract financing for these water-related investments, there is a need to develop demonstrable, bankable, and investment-ready projects of sufficient scale.

Pilot projects can be implemented to evaluate the feasibility and identify opportunities for improvement before the development of a full-scale project. The objectives of such pilot projects include the following:

(i) **Managing risks**. Whether the project is implementing a new technology or a new process, risk plays a major factor in whether business stakeholders will move forward with the proposed change. The pilot project can be used as an opportunity to implement the solution in a limited capacity where the impact of failure is limited. Once the pilot project is executed, the risks that were identified at the beginning of the project can be evaluated in terms of the actual solution being implemented.

(ii) **Validating benefits**. While risk falls on the cost side of the equation, a project would not be considered unless it had some reasonable perceived benefit. A pilot project provides an opportunity to discover or validate benefits by applying the solution concepts in a limited-scope fashion.

(iii) **Promoting change management**. Successful implementation of infrastructure projects often faces challenges because of the following key reasons:

    (a) Differences in approaches to the implementation of different types and levels of projects, programs, and portfolios of projects.

    (b) Integration processes.

    (c) The use of different nonstandardized methodologies and standards creates the problem of their adaptation in the planning and implementation of infrastructure projects.

While the possibilities for pilot projects are extensive, certain principles should be adhered to when selecting and creating them. These principles include the following:

(i) **Urgency of needs**. Different segments of a river basin may face varying levels of urgency in addressing specific issues. Governments and stakeholders should assess which concerns demand the most immediate attention.

(ii) **Existing legal and regulatory frameworks**. The likelihood of success for certain pilot projects—especially those intended for scaling up—depends on the supportive nature of the regulatory framework. Evaluation of legal structures is crucial.

(iii) **Institutional setup**. The sustainability of solutions often hinges on the strength of institutional frameworks and arrangements. It is essential to ensure that proposed solutions are supported by a robust institutional setup to enhance long-term benefits.

(iv) **Extent of capacity building required**. Successful and sustainable schemes necessitate knowledge transfer, capacity building, and effective outreach. A thorough assessment is required to gauge the extent of capacity building necessary for solution success, addressing any technical challenges that may arise.

Based on the previous objectives and principles, some examples of pilot projects that could be implemented in Southeast Asia and the GMS using innovative financing mechanisms could include nutrient credit trading using constructed wetlands on farms, the development of a watershed payment scheme, or leveraging blended finance for the development of wastewater treatment plants.

In conclusion, pilot projects will continue to play a pivotal role in catalyzing the implementation of innovative solutions to combat river pollution in the region. They serve as a crucial platform for identifying potential weaknesses in proposed solutions providing an opportunity for refinement and enhancement. These pilot projects also act as practical tests, offering valuable insights into the feasibility of achieving desired impacts and objectives. Overall, the knowledge gained from pilot projects is expected to be instrumental in assessing scalability and facilitating the transformation of small-scale initiatives into bankable projects that can generate significant impact on river pollution reduction and control in Southeast Asia and the GMS region.

5.3 Support from ADB in Developing River Pilot Projects

As the climate bank of Asia and the Pacific, ADB has been taking the lead in strengthening its climate efforts with several initiatives including the upcoming Climate Tech Hub in the Republic of Korea, the Nature Solutions Finance Hub (NSF Hub), and the ASEAN Catalytic Green Finance Facility (ACGF), among others. The pilot projects described are the perfect example of the type of collaboration and technical assistance support that ADB can provide to regional governments and agencies to assess feasibility solutions for river basin pollution and help develop full-scale projects that will help address nature loss in the region.

The Nature Solutions Finance Hub for Asia and the Pacific

ADB launched a Nature Solutions Finance Hub (NSF Hub) for Asia and the Pacific at COP28, which aims to attract at least $2 billion into investment programs that incorporate nature-based solutions (NBS) particularly focused on capital markets and other sources of private capital.

The hub focuses on catalyzing finance into NBS to support nature-based climate actions and biodiversity conservation. Its goal is to increase investments in NBS projects across Asia and the Pacific by developing scalable and bankable demonstration projects by creating a platform with practical NBS solutions centered around innovative finance approaches. It aims to do this through a partnership approach with global development entities in the nature space along with concessional fund sources who can work together with ADB under a focused roadmap of activities.[253]

The ASEAN Catalytic Green Finance Facility

The ACGF is a $1.9 billion permanent facility under the ASEAN Infrastructure Fund, dedicated to accelerating green infrastructure investments in Southeast Asia. It supports ASEAN member governments to prepare and finance infrastructure projects that promote sustainability and contribute to climate goals. The ACGF is owned by ASEAN member governments and ADB and is administered by ADB.

The ACGF focuses on catalytic green and blue finance and leverages its resources to de-risk infrastructure projects to make them more bankable and attract various sources of funding, including private, institutional, and commercial funds. The ACGF offers access to (i) funds from the ASEAN Infrastructure Fund through two lending products with differentiated pricing based on the differing socioeconomic conditions of ASEAN countries, (ii) financing from nine ACGF partners, and (iii) technical assistance from ADB and ACGF partners.

Technical assistance from the ACGF is provided to enhance capacity to support the origination and structuring of green and blue infrastructure projects. The goal is to develop a pipeline of projects that integrate innovative financing mechanisms that can demonstrate the potential to raise capital at scale. The technical assistance supports upstream studies that identify viable projects, develops and pilots innovative financing mechanisms for early stage project concepts, and helps structure finance-ready projects.[254]

---

[253] ADB. Nature Solutions Finance Hub for Asia and the Pacific.
[254] ADB. ASEAN Catalytic Green Finance Facility.

Appendixes

Appendix 1: Key Features of Major Rivers in Southeast Asia

Mekong River

(i) Geological features

The Mekong River is the longest in Southeast Asia.[1] As the world's 12th longest river, the Mekong River originates about 5,200 meters above sea level at the Tibetan Plateau, flows through the Himalayas in the People's Republic of China (PRC), and onward for almost 4,800 kilometers (km) through Myanmar, Lao People's Democratic Republic (Lao PDR), Thailand, and Cambodia before discharging along the southern coast of Viet Nam.[2]

The Mekong basin is the world's 21st largest at 810,000 square kilometers (km²), with 70% of the basin residing downstream of the PRC–Lao PDR border in a region known as the Lower Mekong Basin (LMB).[3] The LMB extends from the border of Yunnan, PRC to the Mekong Delta, and includes portions of Cambodia, Lao PDR, Thailand, and Viet Nam. The LMB is characterized by fairly flat topography, extensive floodplains, and significant tributaries such as the 3S Rivers (Sesan, Srepok, and Sekong) and Tonle Sap systems.

(ii) Biodiversity highlights

Being biodiversity-rich and ecologically sensitive, the LMB is home to the majority of the Mekong's estimated 1,200 fish species and hosts ecological zones of national, regional, and international importance such as wetlands that have been accorded Ramsar site status (because of their international significance).[4]

(iii) Economic highlights

Countries in the Mekong region recorded impressive economic growth in the last decades, mainly because of the economic opportunities provided by the Mekong River. The annual economic value of water-related sectors in the Mekong River Basin is estimated at almost $35 billion, excluding forestry and tourism. Its rich natural resources support the livelihoods of an estimated 80% of the 65 million people residing within the LMB.[5] Key industries reliant on clean waters include the following:

(a) **Agriculture and fisheries.** The Mekong River provides irrigation, with agriculture using an estimated 70% of the Mekong River water resources. These two sectors are especially significant as they provide substantial foreign currency earnings while bolstering domestic food security.[6] For example, the Mekong River supports the world's largest inland fishery with an annual turnover of close to $4 billion.[7]

---

[1] Mekong River Commission (MRC). Geography.

[2] MRC. 2019. *State of the Basin Report 2018*; OECD et al. 2020. *Innovation for Water Infrastructure Development in the Mekong Region. The Development Dimension*. OECD Publishing, Paris; Effective 1 February 2021, ADB placed a temporary hold on sovereign project disbursements and new contracts in Myanmar.

[3] MRC. Geographic Regions.

[4] Hortle et al. 2015. *Fisheries Habitat and Yield in the Lower Mekong Basin*. MRC Technical Paper No. 47. MRC.

[5] MRC. Mekong Basin.

[6] N. Axelrod. 2016. *The Role of the Mekong River in the Economy*. World Wildlife Fund (WWF).

[7] ADB. 2015. *Investing in Natural Capital for a Sustainable Future in the Greater Mekong Subregion*.

Unsustainable and uncoordinated development is pushing the river system to the brink. As such, there is a strong impetus for governments, businesses, and communities in the region to come together to better manage the river in a way that respects ecosystem limits to support continued economic growth.

Hong River

(i) Geological features

The Hong River—also known as the Red River—is the largest in northern Viet Nam, covering Ha Noi and 25 other provinces. From its source in central Yunnan Province, southwestern PRC, the river flows southeast in a deep, narrow gorge across the Tonkin region, through Ha Noi, and enters the Gulf of Tonkin after a course of 1,200 km.[8]

The Hong River basin can be divided into four subbasins:

(a)   Yuanjiang–Thao River subbasin (Thao River)

(b)   Da River sub basin (Da River)

(c)   Lo River sub basin (Lo River)

(d)   Red River Delta sub basin

Specifically, the Hong River basin has a drainage area of 156,500 km^2, of which 50.3% is in Viet Nam, 48.8% is in the PRC, and 0.9% is in the Lao PDR.[9]

(ii) Biodiversity highlights

The Red River Delta coastal areas are biodiversity-rich and play ecological roles of global importance. They are home to a complex system of various vegetation types, including salt marshes, dunes, and mangroves. In particular, the mangrove and intertidal habitats of the delta form wetlands of high biodiversity, with the Xuan Thy and Tien Hai districts housing the Red River Delta Biosphere Reserve that serves as a globally important site for migratory birds. Relatedly, 78 species of water birds have been recorded in the delta, with 11 of these falling under the category of threatened or near-threatened species.[10]

(iii) Economic highlights

The Hong River has a population of 26 million people and an agricultural area of nearly 1.09 million hectares.[11] The Red River Delta is significant in the following ways:

(a)   It is an important strategic region in terms of politics, economy, society, culture, and the environment.

(b)   It is also significant in the areas of defense, security, and sovereignty.

(c)   It functions as an important bridge in international economic and cultural exchanges, chiefly as the northern gateway of Viet Nam and the Association of Southeast Asian Nations to the PRC.[12]

---

[8]   Britannica. Red River.

[9]   W. Zhou and X. Li. 2022. Fish Fauna of the Red River, Southeast Asia: Indictors and Implications for Planning Fish Species Preserves. *Ecological Indicators*. 141.

[10]   UNESCO. 2018. *Red River Delta Biosphere Reserve, Vietnam.*

[11]   *Vietnam Net.* 2014. Expert says Hong River flow vital for Delta region. 13 February.

[12]   *Vietnam Law and Legal Forum.* 2023. Red River delta asked to lead Vietnam's economic restructuring. 13 February.

Chao Phraya River

(i) Geological features

The Chao Phraya River is situated between Ayutthaya and the Gulf of Thailand and flows south through Thailand's fertile central plain for over 365 km.[13] The river basin in Thailand covers an area of 159,000 km^2, corresponding to 30% of the total area of the country.[14] The basin can be divided into eight sub-basins: Ping, Wang, Yom, Nan, Sakae Krang, Pasak, Tha Chin, and the Chao Phraya mainstream.

(ii) Biodiversity highlights

The Chao Phraya River hosts about 280 fish species including three of the largest freshwater fish in the world, the critically endangered giant barb, giant pangasius, and giant freshwater stingray. However, pollution contributes to extensive habitat destruction that threatens basin biodiversity.[15]

(iii) Economic characteristics

The basin is home to 57% of Thailand's population, particularly in the densely populated Bangkok metropolitan area.[16] It has been significantly used for commerce, has been a center of rice production, and supports 66% of the nation's gross domestic product.

Citarum River

(i) Geological features

The Citarum River is the longest in the Indonesian West Java province, stretching 297 km before reaching the Java Sea, and covers an area of 13,000 km^2.[17] The Citarum River is the source of water for the Jatiluhur Reservoir, which is the country's largest reservoir at 3 billion cubic meters of storage capacity.

(ii) Biodiversity highlights

While the basin is biodiverse and contains 160 plant species, it is also home to a sizeable number of mammals, birds, reptiles, and fish species that have been classified as threatened.[18]

(iii) Economic characteristics

Over 5 million people live in the river basin today, and the river provides water, electricity, and irrigation to more than 25 million people. Specifically, the river provides for people and industries in the following ways:

   (a)   It supplies clean water for Bandung and the capital Jakarta.

   (b)   Provides 80% of the water supply for the capital.

   (c)   Irrigates 400,000 hectares of rice fields.

   (d)   Is a source of energy for three hydroelectric power stations serving three cities.[19]

---

[13]   K. Olson and W. Kreznor. 2021. Managing the Chao Phraya River and Delta in Bangkok, Thailand: Flood Control, Navigation and Land Subsidence Mitigation. *Open Journal of Soil Science.* 11(4).

[14]   GEF IW: LEARN. Flood and Drought Management Tools Chao Phraya Basin.

[15]   WWF. River Stories.

[16]   UNEP. 2016. *A Snapshot of the World's Water Quality: Towards a Global Assessment.*

[17]   Clean Currents Coalition. n.d. *Citarum River, Indonesia.*

[18]   Government of Indonesia, Ministry of Forestry. 2013. *Integrated Citarum Water Resource Management Investment Program.*

[19]   D. Tarahita and M. Rakhmat. 2018. Indonesia's Citarum: The World's Most Polluted River. *The Diplomat.* 28 April.

Unfortunately, the Citarum River is often referred to as the world's most polluted river. Every day, no less than 20,000 tons of waste and 340,000 tons of wastewater—mostly from 2,000 textile factories—are disposed of directly into the once clear and pristine waterways of the Citarum River. Cleaning up the Citarum River and its 22 streams has been classified as a national priority by the Indonesian government, which in 2010 launched a 15-year project to rehabilitate the river.[20] The program is supported by the International Monetary Fund and ADB, which in 2009 had already committed to provide $500 million to fund the rehabilitation of the river.

---

[20]   Cita-Citarum. 2012. *Indonesian lives risked on "world's most polluted" river.*

Appendix 2: Pollution Recovery Highlights of Major European Rivers

Danube River

(i) General Background

Flowing 2,857 kilometers (km) from the Black Forest of Germany to the Danube Delta in Romania and Ukraine, and into the Black Sea, the Danube is the only major European river that flows west to east, from Central to Eastern Europe.[1] While the mainstream Danube flows through 10 countries,[2] the Danube River Basin is the second-largest in Europe, with a total area of 801,500 square kilometers (km²) encompassing 19 countries.[3]

(ii) History of River Pollution and Recovery Highlights

The need to urgently address the polluted waters of the Danube Basin was apparent by the mid-1980s. Population growth coupled with industrialization severely degraded the basin's waters. About 80% of the Danube wetlands and floodplains had disappeared, and nutrient pollution turned the western Black Sea (where the Danube discharges) into a dead zone as oxygen levels were too low to support the survival of most organisms. However, the Danube Basin and the Black Sea are both exhibiting clear signs of recovery from the pollution that has left scars across the region. For instance, from a basin-wide perspective, nitrogen emissions decreased by 20% and phosphorus by almost 50% over the first 15 years of the 2000s.[4]

Rhine River

(i) General Background

The Rhine River is Europe's most important commercial waterway, channeling the flow of trade among France, Germany, the Netherlands, and Switzerland.[5] The 1,320 km Rhine River—which spreads over an area of 225,000 km²— is a classic example of a "multipurpose" river, used simultaneously for transportation, industry and agriculture, urban drinking and sanitation needs, hydroelectric production, and recreation.[6] The environmental problems of the Rhine are significant because it is densely populated (50 million people live along its borders), highly industrialized (10% of global chemical production), and relatively short.

(ii) History of River Pollution and Recovery Highlights

The Rhine was on the brink of dying in the late 1960s when industrial waste turned the river into a turgid, chemical-reeking brew that led to the extinction of the Rhine salmon and the deaths of even the hardiest species of fish and eel.[7] However, conditions in the Rhine and its tributaries have improved markedly. Measures to improve water quality, restore river continuity, and renaturation measures have resulted in the significant recovery of the Rhine biotic communities since 1990. Almost all the Rhine fish species have returned, including migratory fish such as salmon.[8]

---

[1]   WWF. The Danube.

[2]   Britannica. Danube River.

[3]   International Commission for the Protection of the Danube River. Danube River Basin.

[4]   Global Environment Facility. 2016. *Reviving the Danube.* 1 May.

[5]   Cioc. 2006. *The Rhine an Eco-Biography, 1815–2000.* University of Washington Press.

[6]   Brenner et al. 2002. *The Present Status of the River Rhine with Special Emphasis on Fisheries Development.* Food and Agriculture Organization of the United Nations (FAO).

[7]   C. Whitney. 1977. Father Rhine Fallen to Industrial Waste. *The New York Times.* 13 March.

[8]   International Commission for the Protection of the Rhine. 2021. *How Is the Rhine Today?*

Elbe River

(i) General Background

The Elbe River Basin covers an area of 148,000 km², making it the fourth largest-river basin in Central Europe after the Danube, Vistula, and Rhine basins. The total length of the Elbe River from its source in the Krkonoše Mountains to the North Sea is 1,094 km.[9] Inhabited by about 25 million people, more than 99% of the river basin is in Germany and the Czechia, with less than 1% in Austria and Poland.[10] As one of the major waterways in Central Europe, it facilitates the transportation of goods and people, fostering trade and economic development.[11]

(ii) History of River Pollution and Recovery Highlights

By the end of the 1980s, the Elbe was one of the most polluted rivers in Europe and considered doomed. The waters carried a toxic mix of pollutants including nitrogen, phosphorous, mercury, and pentachlorophenol (a highly toxic chemical compound causing fish in the river to develop unnatural growths).[12] However, the Elbe water quality has markedly improved. Average concentrations of heavy metals, organic substances, and nutrients have declined considerably. This has prompted a notable increase in the number of fish species residing in the waters.[13]

Seine River

(i) General Background

The Seine River is an iconic waterway that holds great historical and cultural significance for France.[14] With a length of 754 km, it originates near Dijon, flows through Paris, crosses several important urbanized areas, and discharges into the English Channel. The basin drainage area is about 75,976 km². The Seine River Basin has a population of 16.5 million people and is primarily used for agriculture.[15] It also serves as a vital waterway for trade, transportation, and tourism.

(ii) History of River Pollution and Recovery Highlights

Historically, upstream industrial sewage and municipal waste from a burgeoning population have plagued Seine water quality. Pollution was so severe that swimming in the river was banned in 1923, and by the 1960s, only three species of fish were recorded in Paris. However, improvements in water quality over the past 20 years have led to the reintroduction of swimming in the waters, and an increase in the number and size of fish.[16]

---

9   Flussgebietsgemeinschaft Elbe. The Elbe River Basin.

10   International Commission for the Protection of the Elbe River (ICPER). *The Elbe.*

11   World Atlas. Elbe River.

12   M. Lüpke. 2013. *Rescued Rivers. DW News.* 8 December.

13   ICPER. Water Quality.

14   D. Patel. 2021. *The Seine River Basin. ArcGISStoryMaps.* 9 September.

15   European Commission. Knowledge Hub for Water. 2020. *FACT SHEET: Seine River Basin.*

16   *BBC.* 2023. Paris to bring back swimming in Seine after 100 years. 25 July.

Appendix 3: Pollution Status Mapping—Areas Surveyed by River Basin

River	**Mekong**
Data source	Mekong River Commission (2022). Lower Mekong Water Quality Monitoring Report (2019)
Area(s)	Lower Mekong Basin (LMB): Mainstream and tributaries (Bassac, Nam Ou, Mae Kok, Nam Mun, Houay Mak Hiao, Tonle Sap, Se San, and Sre Pok rivers)
Survey Year(s)	2019

River	**Hong (Red)**
Data source	McGowan, S.; Salgado, J. (2022). Water chemistry from the Red River Delta, Vietnam, 2018 to 2020. NERC EDS Environmental Information Data Centre.
Area(s)	Mainstream and a distributary (Day River)
Survey Year(s)	2019

River	**Chao Phraya**
Data source	Pollution Control Department (PCD), Ministry of Natural Resource and Environment
Area(s)	Mainstream and a distributary (Day River)
Survey Year(s)	2019
Data source (B)	Tomioka et al_2021_JWH_Detection of Potentially Pathogenic Arcobacter supp. in Bangkok Canals and the Chao Phraya River ; Ramadhiani and Suharyanto_2020_EES_ Analysis of River Water Quality and Pollution Control Strategies in the Upper Citarum River
Area(s)	Bangkok
Survey Year(s)	2017, 2018

River	**Citarum**
Data source (A)	West Java Provincial Database (Citarum River)
Area(s)	Citarum (Upper and Lower)
Survey Year(s)	2023
Data source (B)	Ramadhiani and Suharyanto_2020_EES_Analysis of River Water Quality and Pollution Control Strategies in the Upper Citarum River
Area(s)	Citarum (Upper)
Survey Year(s)	2019
Data source (C)	Jubaedah et al_2021_E3S_Status of Water Quality and Level of Trophic in Juanda Reservoir of Purwakarta Regency, West Java, Indonesia
Area(s)	Juanda Reservoir, Citarum Basin
Survey Year(s)	2020
Data source (D)	Sholeh et al_2018_AIPCP_Analysis of Citarum River Pollution Indicator Using Chemical, Physical, and Bacteriological Methods
Area(s)	Citarum (Lower)
Survey Year(s)	2017
Data source (E)	Yokosawa and Mizunoya_2022_S_Improving Water Quality in the Citarum River through Economic Policy Approaches
Area(s)	Citarum (Lower)
Survey Year(s)	2015

River	Danube
Data source	International Commission for the Protection of the Danube River (ICPDR) Joint Danube Survey 4 Database
Area(s)	Mainstream
Survey Year(s)	2019

River	Rhine
Data source	International Commission for the Protection of the Rhine (ICPR)—Water Quality Database
Area(s)	Mainstream
Survey Year(s)	2019

River	Elbe
Data source	UN GEMStat—Water Quality Data Portal
Area(s)	Basin (Germany)
Survey Year(s)	2019

River	Seine
Data source	UN GEMStat—Water Quality Data Portal
Area(s)	Basin
Survey Year(s)	2015
Data source (B)	Yan et al_2022_WR_Unravelling Nutrient Fate and Carbon Dioxide Concentrations in the Reservoirs of the Seine Basin Using a Modeling Approach
Area(s)	Seine (Upper)
Survey Year(s)	2019, 2020

Appendix 4: Addressing Pollution through Policy Measures

Payment for Ecosystem Services

Table A4.1: Four Principal Groups Involved in a Payment for Ecosystem Services Scheme

Key actors involved in PES schemes	Description	Sub-category	Description
1. Buyers	Beneficiaries of ecosystem services who are willing to pay for them to be safeguarded, enhanced, or restored.	Primary buyers	Private organizations and individuals who benefit directly from, and pay directly for, improved ecosystem service provision (e.g., reduced flood risk, clean water, recreational access).
		Secondary buyers	Organizations that buy improved ecosystem service provision on behalf of sections of the public. Secondary buyers can include water utilities, insurance companies, NGOs.
		Tertiary buyers	Buyers who purchase improved ecosystem service provision on behalf of the wider public, i.e., the government.
2. Sellers	Land and resource managers whose actions can potentially secure the supply of the beneficial service	• Farmers • Agribusinesses • Institutional landowners • Large estates • Pension funds • Environmental organizations • Shoreline owners and management authorities	Sellers can be individual landowners, resource managers, or organized groups acting collectively
3. Intermediaries	Serve as agents linking buyers and sellers and can help with scheme design and implementation. Can perform a variety of tasks including • helping sellers assess an ecosystem service "product" and its value to prospective buyers; • introducing buyers and sellers and building rapport between them; • establishing ecosystem service baselines and the scope for additionality; • identifying specific resource management interventions that will deliver service provision; • aggregating multiple landowners and/or managers for more complex schemes; • assisting in determining prices, accessing grants, structuring agreements, and agreeing on a mutually acceptable payment regime; • activities related to implementation (including monitoring, certification, verification); and overall scheme administration.		
4. Knowledge Providers	Include scientists researching ecosystem service provision, resource management experts, valuation specialists, land use planners, regulators, and business and legal advisors who can provide knowledge essential to scheme development.		

NGO = nongovernment organization, PES = payment for ecosystem services.
Source: E. Fripp. 2014. Payments for Ecosystem Services (PES): A practical guide to assessing the feasibility of PES projects. CIFOR.

Payment for Ecosystem Services

Table A4.2: Examples of Extended Producer Responsibility Schemes Implemented Globally

Country		
European Union (EU)	All member states have product takeback systems. The framework is established through the EU, but operational aspects are advised by states.	Four main types in all states: packaging, batteries, end-of-life vehicles, and waste electrical and electronic equipment. Some states also have different material lists.
United States	There is no national extended producer responsibility (EPR) policy. Individual states develop and implement their policy. Today there are 89 EPR laws in 33 US states.	A wide range of materials.
Canada	A Canada-wide Action Plan for EPR occurs at the province or territory level. There are more than 30 federal and provincial producer stewardship programs.	A wide range of materials.
People's Republic of China	The new EPR policy was introduced in 2016–17 by the PRC State Council, and relevant laws and regulations are to be formed by 2025.	Certain materials: electrical products, batteries, vehicles.
Japan	Home Appliance Recycling Act.	A wide range of materials including construction and demolition waste.
Republic of Korea	Resource Saving and Recycling Promotion Act 1992, Resource Circulation of Electrical and Electronic Equipment and Vehicles 2008.	Household and industrial materials.

EPR = extended producer responsibility.

Source: Literature review, EY analysis.

Table A4.3: Stakeholder Roles in Extended Producer Responsibility

Stakeholder	Role(s)
National Government	Setting the policy and legislative framework, including the following: • Defining the producers and products concerned. • Setting the actual responsibilities for the producers, e.g., quantified targets for takeback, collection, and recycling of waste. • Defining the roles of other actions, e.g., local municipalities, and the informal waste sector. Dialog with these sectors is important. • Accreditation and/or approval and monitoring of EPR schemes to ensure good functioning and enforce compliance. • Taking steps to combat illegal imports of packaging or packaging waste.
Local Municipalities	Typically, responsible for waste collection from households and businesses. Including providing readily accessible infrastructure. Provision of information to the public.
Producers and Businesses (including manufacturers, consumer goods companies, and retailers and/or distributors)	Responsible for meetings and targets set by the government, including • Creation of EPR schemes including setting up non profit or for-profit producer responsibility organizations (PROs) in the case of collective EPR. • Administering and running EPR schemes, which may include setting registration and product fees, collecting fees (typically from consumer good companies), relationships with waste collectors and processors, reporting collection and recycling rates, and possibly take-back of waste packaging (retailers and/or distributors). • Paying fees to EPR schemes based on the packaging material they place on the market. • Providing information to producers and consumers on how to use EPR schemes.
Waste Management Companies	Collection and management of waste through contracts with local municipalities, PROs, or individual producers. Should receive funds from EPR schemes for handling packaging waste.
Informal Sector (e.g., waste pickers)	Any existing informal sector actors should be allowed to participate in EPR schemes, for example, by contributing to the effective collection of recyclable waste.
Consumers/Citizens/Households	Returning waste products at the end of their useful life, using the infrastructure provided.

EPR = extended producer responsibility, PRO = producer responsibility organization.

Source: WWF and Institute for European Environmental Policy. 2020. *How to Implement Extended Producer Responsibility (EPR)*. WWF.

Appendix 5: Key Learnings from the Knowledge Roundtable on River Basin Pollution

Through the Zero Source Pollution Initiative, the Asian Development Bank (ADB) and its partners aim to prepare high-impact, replicable projects and mobilize $1 billion in investment by 2030 to address the issue of water pollution caused by waste. As part of this initiative, ADB hosted a knowledge roundtable at the Anantara Siam Bangkok Hotel on 2 November 2023. The event featured presentations and panel discussions with participants from various organizations. The sessions covered were as follows:

(i)   Session I: The Impact of River Basin Pollution in Southeast Asia

(ii)   Session II: Meeting the Challenges: Greater Mekong Subregion Perspectives on Technical, Institutional, and Governance Issues

(iii)   Session III: Emerging Global Examples

(iv)   Session IV: The Need for Innovative Financing Solutions

(v)   Session V: ADB Technical Assistance and City Discussions

Based on the various presentations and panel discussions, several key learnings emerged that were not initially included in this report. Therefore, Appendix 5 serves as an addition to the report, highlighting information and knowledge that was gained from the roundtable. The authors have aligned additional learnings in this appendix with the structure of the main report. The key learnings from the roundtables can be found below.

3.2 The Impact of Pollution on Climate Change

- **Climate change is exacerbating the negative impacts of pollution**. Because of climate change, the problems stemming from pollution persist for weeks instead of just days. What was once ankle-deep wastewater has now deepened to waist-deep levels.

4.0 Addressing Pollution: Global and Regional Insights

- **The Asian Institute of Technology has developed innovative solutions in plastic monitoring, wastewater management, and groundwater governance**. In collaboration with the Bill and Melinda Gates Foundation, the Asian Institute of Technology (AIT) has established a Water Sanitation Center, contributing significantly to the development of innovative solutions. AIT expertise extends to water resource management, quantity, and quality, addressing critical aspects of water sustainability. Their Digital Solutions for Plastic Litter Monitoring—implemented in the Mekong River—utilizes an open, scalable, platform-based approach with real-time analytics and low-cost closed-circuit television modules, employing artificial intelligence to trace the source of plastic pollution. The AIT commitment to inclusive wastewater and fecal sludge management is evident in their City-wide Inclusive Wastewater and Fecal Sludge Management Plan, supporting local governments in Kratie, Cambodia and Luang Namtha, Lao People's Democratic Republic. AIT also operates a performance testing lab for prefab residential wastewater treatments, holds ISO certification, and aids labs in showcasing their effectiveness to the public and private sectors. Additionally, AIT addresses critical gaps in groundwater governance, developing the Groundwater Governance Framework and assessing the unique challenges of Cambodia, where institutions, policies, and laws related to groundwater management are absent.

4.1 Nature-Based Solutions

- **Nature-based solutions as a form of infrastructure can address biodiversity and climate change simultaneously**. Solutions can take the form of either nature-inspired or nature-derived approaches, but the term "nature-based" specifically refers to enhancing nature beyond its current capacity or level. The International Union for the Conservation of Nature standard serves as the global benchmark for these nature-based solutions. It is crucial to view ecosystems as a form of infrastructure in their own right, as they possess the ability to simultaneously address biodiversity and climate change holistically. Consequently, it is essential to consider both traditional (gray) and natural (green) infrastructure to create effective and sustainable solutions.

4.2.1 Payment for Ecosystem Services

Payment for Ecosystem Services (PES) has been used extensively in the Yangtze and Yellow Rivers. PES has been integrated into the legal framework of the Yangtze and Yellow River regions. The commitment to green financing is exemplified by the allocation of CNY30 billion through government budget transfers, establishing a vertical eco-compensation system. Evolving from this, a significant shift is underway toward horizontal eco- compensation, wherein downstream areas contribute financially to upstream conservation efforts. This approach seeks to balance the ecological impact throughout the entire river system. In a progressive move, plans are underway to implement horizontal eco-compensation across all tributaries of the Yangtze River, reflecting a comprehensive strategy to promote sustainable management and conservation practices along these vital waterways.

4.3 Institutional Arrangements

- **Poor understanding of cross-ministry impacts and weak legal frameworks hampers effective river basin management.** Although river basin management plans exist in the Greater Mekong Subregion, their implementation has been notably poor. This deficiency stems from a siloed approach to projects, a lack of understanding of their broader impacts across various ministries, and has highlighted the imperative for enhanced intragovernmental cooperation. Additionally, a legal gap hampers implementation, particularly concerning the disparity between upstream and downstream countries, indicating a need for legal frameworks to bridge these gaps.

- **International River Foundation has contributed significantly to global river cleaning efforts.** The International River Foundation (IRF) has played a pivotal role in addressing the severe degradation of rivers, considered among the most compromised ecosystems globally. At the forefront of their impactful initiatives is the prestigious International River Prize, recognized as the most esteemed environmental award globally. The inaugural recipient—the River Mersey—set a precedent for outstanding river management and restoration efforts. The International River Symposium—organized by IRF—serves as a crucial platform, bringing together river managers, policymakers, scientists, consultants, and various stakeholders to collaborate on holistic solutions. The Resilient Rivers Blueprint Hub is another noteworthy initiative aimed at enhancing river resilience. With support from CEOs and alumni, IRF has been instrumental in advocating for innovative approaches. An illustrative example is the shift in treatment methods in Brisbane, Australia, where initial onsite treatments were mandatory but later proved less efficient and more expensive at $43,000 per metric ton compared to the offsite approach costing only $248 per metric ton. The IRF commitment to recognizing and promoting effective river management strategies positions them as a significant force in the global conservation and restoration of vital river ecosystems.

- **Australia's Department of Foreign Affairs and Trade requires gender and climate considerations in the projects it funds and frequently explores innovative financing mechanisms.** The Australian government (new in 2022) prioritizes scaling up efforts in climate, energy, and water sectors, recognizing the imperative for a comprehensive approach. Despite having a significant budget for funding purposes, the Department of Foreign Affairs and Trade (DFAT) mandates that 80% of projects integrate meaningful considerations for climate change and gender equality. Acknowledging the limitations of traditional grant funding, DFAT actively explores diverse mechanisms like blended finance to bolster their capabilities.

- **The Department of Foreign Affairs and Trade is committed to supporting the Greater Mekong Subregion in its water resource challenges.** The Mekong–Australia Partnership and Australia's Partnerships for Infrastructure initiative underscore DFAT's commitment to fostering collaboration and innovative approaches in addressing water resource challenges in the region. Additionally, DFAT engages in bilateral agreements, particularly through the Greater Mekong Water Resources Programme, which now extends its coverage to include climate change and energy. Given the growing significance of hydropower, DFAT also navigates complexities associated with trust funds of banks, such as the Green Climate Fund.

4.3.1 Intergovernmental Cooperation

- **The International Commission for the Protection of the Danube River plays a crucial role in assisting member countries to comply with the European Union Water Framework Directive.** Encompassing over 800,000 km² and involving 19 countries, the Danube River Basin rivals the size of the Mekong. Established in 1994, the Danube River Protection Convention is dedicated to safeguarding the basin's rivers. The International Commission for the Protection of the Danube River plays a pivotal role in assisting EU member states in complying with the mandatory European Union Water Framework Directive. Challenges faced include historical pollution, climate change, green navigation, and the imperative for transnational monitoring and an accident warning system. Basin-wide management plans—exemplified by the Danube Declaration and Danube River Basin Management Plan—along with a climate change adaptation strategy, are pivotal in tackling these challenges. The European Union Water Framework Directive—employing a river basin approach—underscores international coordination, aiming for good water status, environmental objectives, a no deterioration rule, and cost recovery. The planning cycle—featuring a 6-year review and incorporating public participation—ensures adaptability. The Transnational Monitoring Network evaluates chemical water status, ecological status, and treatment status, demonstrating notable improvement from 2009 to 2018.

- **The International Commission for the Protection of the Rhine has made significant achievements in improving the Rhine's biodiversity.** Despite being smaller than the Mekong, the Rhine River is densely populated, with nine states along its course. Established in 1950, the International Commission for the Protection of the Rhine (ICPR) focuses on water quality, quantity, and biodiversity. While the states collectively agree on overarching goals, they independently determine the methods to achieve them. ICPR has achieved significant milestones, with notable reductions in heavy metals, increased oxygen levels, and a rise in invertebrate species. A remarkable recovery includes the return of 70 native fish species, including the Atlantic salmon, that went extinct after World War II. Floodplains and riverbanks have been restored and water bodies reconnected. The Program Rhine 2040 aims to update climate change strategies and reduce micropollutants. ICPR's success stories include the revival of the Atlantic salmon, serving as a flagship species. This initiative—requiring pristine water quality—showcases the potential to rejuvenate ecosystems by bringing back keystone species like salmon.

4.3.2 Intragovernmental Cooperation

- **Intragovernmental cooperation improves the implementation of national strategies.** Thailand has already developed a national water resource management strategy to safeguard the quality of our water resources. However, to effectively implement this strategy, it is essential to improve collaborative efforts with local authorities by enhancing their capabilities and offering technical support where needed. This collaborative approach is especially critical at the sub basin level, where the intricacies of water management are most pronounced.

- **Water resource management is a complex process that requires a holistic approach.** Southeast Asia is in urgent need of a comprehensive approach to water resource management, moving beyond isolated projects to embrace holistic, cross-sector programs across the region. The expertise of global organizations like ADB and the World Bank is crucial in achieving these goals. Active engagement of all stakeholders—including the private sector—is essential for success. Despite having legislation and water laws in place, the challenge lies in addressing the poor implementation of these measures. A robust monitoring system is necessary for accountability and success, and fostering a culture of information sharing among stakeholders is vital for promoting cooperation and achieving sustainable water management goals in the region.

- **Paris has invested a significant amount to re-enable swimming in the Seine**. The Paris Region Seine-Normandy Water Agency has spearheaded a significant river clean up initiative for the Seine, aiming to restore its water quality to a level where people can once again swim in it. A century ago, swimming in the Seine was a common practice, but by the 1960s, water quality had declined to the extent that swimming became prohibitive. Through the implementation of Protocole d'Engagement and Plan d'Actions Baignade, the agency has engaged numerous stakeholders in the restoration efforts. With a financial commitment of €1.2 billion –1.4 billion, the initiative focuses on upgrading wastewater treatment plants to enhance water quality and make the river suitable for recreational activities like swimming. The success of this endeavor is evidenced by the transformation from a polluted waterway to a revitalized Seine, allowing people to once again enjoy its waters for swimming.

4.4 Financing Mechanisms

- **The International River Foundation has contributed to the development of innovative financing mechanisms for river sustainability projects.** To enhance the financial viability and valuation of river projects, the International River Foundation (IRF) has expanded beyond its traditional avenues of the International River Symposium and International River Prize. Recognizing the limitations of traditional funding methods, IRF actively collaborates with governments and corporate entities to establish blended finance approaches for more sustainable river projects. Through initiatives like the River Resilience Blueprint, incorporating elements such as adaptive management, financial security, institutional arrangements, and a systems approach, IRF aims to strengthen the resilience of river ecosystems. The foundation advocates for the quantification of the value of ecosystems, emphasizing the importance of assessing the iconic value of rivers. The River Prize now mandates that projects demonstrate bankability, aligning with the need for markets dedicated to environmental preservation. Cost-benefit and cost-effective analyses are integral components, prompting a thoughtful consideration of the impact of inaction and underscoring the value of proactive intervention in river management. This comprehensive approach by IRF seeks to create financially sound and ecologically valuable river projects, fostering a more sustainable future for these vital waterways.

4.4.2 Public–Private Partnerships

- **The Ganga River cleanup has benefited greatly from a unique public–private partnership approach.** The Ganga River holds significant economic importance, contributing to over 40% of India's GDP, but has faced significant deterioration over the last few decades. The Ganga Cleanup Initiative—supported by a dedicated task force—has been acknowledged by the United Nations as one of the Top 10 World Restoration Flagships to revive the natural world. The initiative involves a substantial $4.5 billion project, with committed funding of $940 million from banks. The success factors identified include the critical role of institutions in the cleanup process. Moreover, the adoption of a hybrid annuity model in public–private partnership for the Ganga Basin signifies a shift from a construction-centric approach to an outcome-focused strategy, marking a significant step toward sustainable river restoration.

4.4.4 Other Financing Mechanisms

- **Quantifying ecosystem values can attract private investments.** It is imperative to foster a better understanding of the inherent value of ecosystems to increase investments in their preservation. For instance, coral reefs—despite their substantial ecosystem value—suffer greatly. While the private sector may have mixed levels of interest, there is an urgent need to engage them in addressing these critical issues.

- **The land value capture method can be used for river basin projects.** The Fujian Xianyou Mulan River Basin project—approved in August 2022—is a $549 million project involving contributions of $200 million from ADB and $140 million from the China Development Bank. Encompassing an expansive area of 1,700 km², the project addresses various issues including flooding, solid waste, agriculture runoff, and urban wastewater. The project aims to achieve three main outputs: piloting innovative financing mechanisms, strengthening institutional capacity for environmental management, and implementing flood management, ecological restoration, sanitation, and water resources improvement systems. The utilization of the land value capture method in the first output and the promotion of ecosystem protection as a business represent innovative financial approaches to establish a sustainable financing mechanism for environmental management. The lessons learned from this initiative have the potential for replication in other rural areas in the PRC and beyond.

- **The land value capture method is approached with caution in the People's Republic of China.** The concept of land value capture is approached cautiously by PRC governments because of its sensitive nature and potential side effects. The understanding is that a cleaner river not only enhances the overall living environment but also has a positive impact on the land value in urban areas. Although the immediate project might not lead to an immediate increase in land value, the anticipation is that it could contribute to value appreciation in the future, especially when the land is eventually sold. This nuanced approach acknowledges the intricate relationship between environmental improvements, urban development, and the long-term economic valuation of land.

Appendix 6: Knowledge Roundtable Project Development Exercise for Cities

Through the Zero Source Pollution Initiative, the Asian Development Bank (ADB) and its partners aim to prepare high-impact, replicable projects and mobilize $1 billion in investment by 2030 to address the issue of water pollution caused by waste. As part of this initiative, ADB hosted a knowledge roundtable at the Anantara Siam Bangkok Hotel on 2 November 2023. The event included presentations and panel discussions with participants from various organizations. During Session V titled "ADB Technical Assistance and Cities' Discussions," a workout session was held for eight cities that attended the event. The objective of the workout session was to enable ADB to preliminarily identify areas where it can provide technical, financial, or other forms of support to these cities.

Each city discussed the key challenges it was facing from a city context, including environmental issues, pollution, and other shocks and stresses. They also explored possible solutions to these challenges, which encompassed infrastructure projects, non-infrastructure projects, technical assistance, knowledge and technology, and partnerships. Additionally, the cities considered potential financing mechanisms and the next steps required. Each city delivered a presentation based on their discussions, and the next section summarizes the presentations from the eight participating cities.

Roundtable 1: Phibun Mangsahan, Thailand

Phibun Mangsahan—a small municipality in Ubon Thani province—faces environmental challenges because of untreated wastewater affecting the Mun River, impacting agriculture, public health, and biodiversity. The municipality received funding for a solid waste facility and waste-to-energy solutions but struggled to secure central government funding for a crucial wastewater treatment plant. Given its small size, it struggled to garner attention and funding, especially since the district relies less on tourism, with the main impact being on agriculture, public health, and biodiversity. The river pollution is also a transboundary problem.

Several project options were proposed, emphasizing a broader river-based small-town development project covering agriculture, wastewater treatment, and nature-based flood prevention. The municipality considered presenting itself as a pilot city, potentially serving as a role model for similar small cities. Other suggestions included grouping similar cities for sector development, focusing on the Mun River basin, and exploring distributed wastewater management solutions.

Additionally, Phibun Mangsahan highlighted other areas in which they require support, including knowledge of single-use plastic waste management, mentoring from experts, network partnerships for quick problem-solving, securing funding sources, developing a plastic waste management model, and investing in wastewater treatment systems. The municipality sought budget allocations for trials of waste management models, emphasizing the need for financial support to address environmental challenges and promote sustainable development.

Roundtable 2: Hat Yai, Thailand

Hat Yai is grappling with significant wastewater issues. A wastewater system—constructed in the early 2000s with a B1,800 million ($50 million) investment from an environmental fund—now faces deteriorated pipe systems and limited drainage capacity because of the absence of a maintenance fund. Only 50% of the system is operational. The Waste Management Authority has a limited budget when it comes to maintenance, which can be costly, especially for pipes located beneath canals. Hat Yai is using its municipal budget for maintenance, but this is not financially sustainable. Solid waste is also a pressing concern, with about 1.2 million tons still present in the city. Managing solid waste is challenging because of limited landfill space. The Greater Hat Yai Development Council (a public–private partnership in Hat Yai) management has not been able to meet the required standards and has ceased operations. They were given a contract for 250 tons per day, but the actual clearance rate was about only 150 tons daily. Negotiations are now in progress to secure a new contractor.

For wastewater treatment, a feasibility study estimated an annual cost of about B100 million ($3 million) to maintain Hat Yai's wastewater management system. For solid waste, Hat Yai is exploring a potential collaboration with the World Wide Fund for Nature to engage informal waste pickers. This includes implementing increased incentives (higher fees) for plastics and sorted waste, as well as establishing connections between waste pickers, junk shops, and waste markets. The city is in the testing phase for this platform and is also investigating waste sorting methods to reduce the remaining solid waste within the city.

To enhance the financing of the wastewater maintenance system, Hat Yai's strategy involves improving the collection of water taxes from households. Additionally, there are proposals to foster co-investment collaborations between the private sector and the government within the framework of the United Nations Environment Programme. Such partnerships could offer more favorable interest rates compared to private loans. However, the feasibility of this approach hinges on prevailing regulations and the restricted borrowing capacity of local governments. Moreover, an alternative option involves securing funding from both national and local government sources while seeking technical assistance from organizations like ADB and the United Nations.

Roundtable 3: Roi Et and Aranyaprathet, Thailand

Roi Et is confronted with a critical water management issue because of the absence of proper treatment facilities for its primary water receivers: the public pond and the 6 km canal encircling the city. The lack of treatment plants poses a significant challenge as it prevents the purification of water before reaching these crucial sources, necessitating urgent water treatment solutions. Compounding the issue, the city lacks a suitable area for constructing treatment plants.

In response to these challenges, Roi Et is actively seeking technical assistance to devise effective wastewater treatment methods. They are also requesting a rapid assessment to determine the optimal approach for establishing the necessary treatment infrastructure. To tackle wastewater issues comprehensively, Roi Et is considering various projects, including initiatives for support, survey, analysis, and water quality management. Additionally, the city is exploring a moat development project for tourism.

Roi Et is seeking budgetary support from the Department of Public Works and Town and Country Planning. Collaborative efforts are being pursued with entities such as the Japan International Cooperation Agency, the Department of Public Works and Town and Country Planning, the Royal Irrigation Department, and the Office of Natural Resources and Environmental Policy and Planning. Specific needs encompass technical support for wastewater treatment concept design and a rapid assessment (pre-feasibility study) to develop the wastewater treatment plant.

In a similar scenario, Aranyaprathet is grappling with its treatment system, relying on an open pond without a treatment system that requires dredging every 6 months. Situated about 3 km from the city, this open pond has led to soil mixture issues, triggering complaints from Cambodia as water leaks into the canal, ultimately reaching the Tonle Sap River. Aranyaprathet needs technical support to enhance the existing treatment system or develop a water treatment plant. Furthermore, sewerage improvement—particularly in water quality and walkway infrastructure—is a crucial requirement.

Roundtable 4: Surat Thani, Thailand

Surat Thani is home to 130,000 people and has 76,000 households. This has led to the generation of 90 million liters of wastewater per day. There is no wastewater treatment system in place for the city, and the majority of households use septic tanks to manage their wastewater. As Surat Thani is situated near the sea, high tides help wash and clean pollution spots. However, during periods of prolonged low tides, stench and eutrophication become major issues for the city's residents. Surat Thani has a robust water monitoring system with 21 checkpoints throughout the city. These checkpoints range from "excellent" to "degraded." Surat Thani aims to focus on the checkpoints classified as "on the brink of degradation" to prevent them from reaching a degraded status.

To prevent checkpoints on the brink of degradation from deteriorating further, Surat Thani proposes the construction of five wastewater treatment plants. These will include a combination of 125 cubic meters per hour (m^3/hour) (Guntelai and Don Mau) and 20 m^3/hour (Wat Mai, Provincial Stadium, and Family Court) wastewater treatment plants. A preliminary study for the five selected locations has already been completed. These projects would cost about B200 million for the 125 m^3/hour and B50 million for the 20 m^3/hour treatment systems. The next stage involves conducting a feasibility study, but Surat Thani lacks the budget for this purpose.

Surat Thani is looking to the Wastewater Management Authority of Thailand to fund these wastewater treatment systems. However, they are also open to exploring other innovative financing mechanisms that could assist them in achieving their desired goals. Moreover, Surat Thani will require technical support to execute this ambitious project. The timeline for the completion of the entire project is estimated to be roughly 3 years (1 year for the feasibility study and 2 years for the full construction of the five wastewater treatment plants).

Roundtable 5: Chanthabuly District, Lao People's Democratic Republic

The Chanthabuly District in Lao People's Democratic Republic faces six key challenges, including capacity limitations in technical expertise and financial resources, difficulties in financing equipment and maintenance, poor waste collection, inadequate stormwater infrastructure, potential risks from the sewerage system, and scarce landfill space. In response, the district has identified five solutions: public education for waste behavior, building international knowledge networks, increased public financial responsibility, innovation in waste management, and enhanced understanding of waste types and quantities.

To tackle these challenges, the district has devised six strategic steps and financing mechanisms for sustainable development. These steps involve exploring innovative financing and technologies while prioritizing health and safety. Waste management and sanitation are targeted to enhance dignity and attractiveness, with community engagement encouraged. Education is pivotal in raising awareness about tax allocation and empowering the local population. Businesses are urged to collaborate—considering workforce constraints and subsidies—with performance measurement as a key factor. The district aims to ensure infrastructure sustainability through cost analysis and alignment with population growth. Additionally, it addresses tourism concerns by promoting responsible behavior in high-tourist areas and environmental stewardship in transboundary regions.

Roundtable 6: Vang Vieng, Lao People's Democratic Republic

Vang Vieng is a highly popular city for tourism, attracting about 700,000 visitors annually. Visitor numbers are expected to continue to rise in the coming years, which is likely to impact the local water supply given the level of infrastructure. Only 65% of the downtown area is serviced by a piped water supply; other areas rely on groundwater or the Nam Song River for their water supply. Although there are not many industrial facilities along the Nam Song River (the primary river running through the city), many hotels and businesses discharge their wastewater into this river, causing pollution and biodiversity loss issues. Some hotels have septic tanks, but the treatment quality of these tanks is inadequate.

To secure the future of Vang Vieng's water supply, the government aims to enhance its groundwater quality monitoring and assessment capabilities and better manage the wastewater entering the Nam Song River. The government lacks both the technical and financial capacity to support the development of infrastructure necessary to achieve these objectives. While there are capabilities for collecting river quality data, the ability to assess groundwater quality is limited. Similarly, the volume of wastewater entering the river because of the absence of wastewater treatment facilities is unsustainable.

The government has not yet conducted any preliminary studies on the possibility of a groundwater quality monitoring system. However, the government has already carried out several preliminary studies on the construction of wastewater treatment plants but lacks the budget to proceed with a feasibility study. ADB has previously supported the Vang Vieng government in developing new roads to promote tourism, and the Vang Vieng government hopes that they will extend their support to the development of groundwater quality monitoring systems and wastewater treatment facilities.

Roundtable 7: Binh Duong, Viet Nam

The Binh Duong province in Viet Nam is poised for economic growth of 8%–8.5% and is a key player in the Southern Key Economic Region, attracting considerable foreign investment. With over 30 industrial zones, it ranks second only to Ho Chi Minh City in foreign direct investment. However, the province faces challenges with its municipal wastewater treatment infrastructure, leading to a low household connection rate of only 50%. The minimum wastewater treatment fee linked to water tariffs further discourages residents from connecting. Environmental issues like urban flooding, surface water pollution, and illegal waste discharge persist because of limited financial capacity and low awareness hindering infrastructure development.

The ongoing Binh Duong Water Environment Improvement Project (2023–2028) seeks an estimated total investment of $310.80 million, with a $230.76 million contribution from the World Bank (numbers have been rounded). The project includes two planned wastewater treatment plants (WWTPs): Ben Cat WWTP and Tan Uyen 2 WWTP. Plans are underway to consolidate existing separate WWTPs (Thu Dua Mot, Thuan An, Di An, and Tan Uyen) into two (Ben Cat and Tan Uyen 2) to enhance treatment quality and reduce pollution. It is unclear if these plants fall under the World Bank-financed initiative. The Binh Duong project management unit—responsible for implementation—faces institutional and financial limitations, seeking support to achieve wastewater treatment goals.

With collaboration in the ongoing project as a potential funding source, further partnerships with international organizations, government bodies, and private investors could support the proposed project. Successful capital recovery strategies and innovative economic models may attract investment from various stakeholders.

Roundtable 8: Baguio City, Philippines

The economy of Baguio City is primarily driven by tourism. Tourism revenues significantly contribute to the local economy, and the tourism industry is one of the largest employers since most businesses in the city operate within this sector. However, the city is facing several challenges regarding water management and wastewater treatment.

Baguio City predominantly manages its wastewater using septic tanks, with about 90% of residents serviced through fecal sludge treatment plants with a capacity of about 20 m^3. Communal septic tanks are relatively rare, covering less than 1% of Baguio residents. The sewer network only covers about 10% of the city's residents, with wastewater treatment plants having a capacity of about 8,600 m^3/day.

The local government aims to expand the sewer network, but this task is challenging because of the city's mountainous topography. The hilly terrain has also led to issues with accessibility to houses with septic tanks, with some houses not being accessed for nearly 20 years. The fee collection system is inadequate, as sanitation fees are not universally collected. The local government collects very minimal sanitation fees because they are only collected from a small percentage of the population: the households that avail of the desludging service. This situation is less than ideal, given that onsite sanitation services for septic tanks are estimated to be seven times more expensive than sewerage services in terms of monthly household expenditures. However, a lack of budget hinders the provision of universal and efficient sanitation services.

Nonetheless, Baguio City has developed the Baguio Resilient City Tourism Project, which consists of a two-pronged approach to address its wastewater treatment needs. Output 1 involves an overhaul of urban infrastructure and services, including the construction of a Balili sewage treatment plant with a designed capacity of 12,000 m³/day using biological treatment technology. It also encompasses the rehabilitation of networks, the development of a regional sanitation plan, capacity building and institutional strengthening, and a tariff reform program. Output 2 focuses on enhancing the productivity of tourism workers, involving private sector-led skills and training networks for workers in accommodation, food services, transport, farm tourism, and tour operations. The aim is to develop Baguio into a highly productive tourism industry, as a robust economy is better equipped to fund and support its infrastructure requirements.

ADB is already collaborating with the Baguio City government for the construction of the Balili sewage treatment plant through a $62.4 million loan. Nonetheless, the city government is exploring other forms of funding for various infrastructure needs, including public–private partnerships and the imposition of a tourism or environmental levy.

www.ingramcontent.com/pod-product-compliance
Lightning Source LLC
Chambersburg PA
CBHW061235270326
41929CB00031B/3496